Plants of San Luis Obispo

Their Lives and Stories

Text and Photography by Matt Ritter

KENDALL/HUNT PUBLISHING COMPANY
4050 Westmark Drive   Dubuque, Iowa 52002

This book is dedicated to my spirited daughter, May Sahana Ritter, on her first birthday.

Back cover photo by Wayne Chapman.

Copyright © 2006 by Matt Ritter

ISBN 978-0-7575-2678-0

Kendall/Hunt Publishing Company has the exclusive rights to reproduce this work, to prepare derivative works from this work, to publicly distribute this work, to publicly perform this work and to publicly display this work.

All rights reserved. No part of this publication may be reproduced, stored in a retrieval system, or transmitted, in any form or by any means, electronic, mechanical, photocopying, recording, or otherwise, without the prior written permission of Kendall/Hunt Publishing Company.

Printed in the United States of America
10 9 8 7 6 5 4 3

CONTENTS

Acknowledgments iv

Introduction v

Coastal Scrub & Chaparral... 1

Oak Woodlands & Riparian 37

The Sand Dunes 69

Grasslands & Wildflowers 85

Salt & Freshwater Marsh 115

Weeds 133

Further Reading 146

Glossary.. 147

Index... 151

Acknowledgments

A number of people helped this book come to be. I would like to thank Dirk Walters, V.L. Holland, George Thompson, and Wayne Chapman for helpful information and advice about the manuscript. I am grateful to Charlene Peterson for her creativity and help with the graphic design of the book. I would also like to thank Ben Carter and Cedrick Villaseñor, two students who have accompanied me on many trips into the field and who have taught me far more than I have taught them. I am especially indebted to David Keil for his mentoring, encouragement, friendship, and a very thorough read of the manuscript. I have learned more about the plants of this area from him than any other source. His patience and vast botanical knowledge never cease to amaze. Lastly, I would like to express my appreciation for my wife, Sarah Allen Ritter. If not for her, this book would not exist, and it is only because of her modesty that she is not listed as a co-author. For all her help and kindness I am more grateful than I'll ever be able to say.

INTRODUCTION

This book is a natural history guide to plants in the San Luis Obispo area. There are more than 1,700 different native species within the borders of the county (more than in the entire state of Alaska) and many other introduced weeds and horticultural plants. Although describing all these species is far beyond the scope of this book, I have included photographs and descriptions of plants that are abundant, widespread, and commonly encountered in the area surrounding the city of San Luis Obispo and western portions of San Luis Obispo County. Many of the plants in this book dominate parts of the landscape, whereas others are not dominant, but are clearly visible or especially attractive. Every plant tells a story about a place and the workings of nature, and this book is an attempt to relate a small portion of that story.

I have included photos and descriptions of 206 plant species and referred to various others in the text. These species are in sixty-four different plant families, with most species described in the sunflower (Asteraceae), bean (Fabaceae), lily (Liliaceae), rose (Rosaceae), grass (Poaceae), and mint (Lamiaceae) families. The photographs, which were taken with a small, point-and-shoot digital camera, are intended to help the reader identify plants of interest during hikes and other outdoor activities.

The plants in this book are arranged by the habitat or plant community where they are most likely to be encountered. This is often a difficult task, as natural areas around San Luis Obispo are an intricate mosaic of many different habitats, and many of the dominant plants cross the artificial boundaries we define between plant communities. Some of the most common and widespread non-native species are included in the chapter on weeds. However, I have chosen to describe certain weeds along with the native plants because they occur predominantly in certain plant communities or habitats.

Below each plant photograph is the common name, followed by the Latin name, the person or persons who named the species or genus, and the years that individual or individuals lived. With few exceptions, the common and Latin names of plants used in this book are the same as those in the *Jepson Manual: Higher Plants of California*, which was published in 1993, but still stands as the standard by which California plants are named. In some instances new information has led to a recent name change, and I have used those new names in this book. On the third line below the photograph is the name of the plant family to which the species belongs, followed by a suggested pronunciation for the Latin name. Pronunciation guides represent one way to pronounce the scientific name, but certainly not the only way, as no distinctly correct way exists. Within the pronunciation guides, capitals indicate the syllable of emphasis.

These introductory lines are followed by a short non-technical description of the plant. Every attempt has been made to minimize the use of technical botanical terms, and those remaining in the descriptions are defined in the glossary. Interested readers can find additional details in other publications referenced in the further reading section of the book.

Each discussion contains various tidbits of interesting information about the plant or its close relatives. These may include anatomical features, ecological information, edibility or toxicity, how the plant was or is used by humans, horticultural information, how the flowers are pollinated, the meaning behind the scientific names, or other noteworthy tales related to the species. The descriptions of a plant's food and medicinal uses are for interest only and are not intended to encourage the reader to try any of these. Many of the species described are completely or partly poisonous, and plants should always be properly identified before being used or eaten.

I hope this book will aid interested readers on their journey to learn more about the plants around us. Knowledge of the names, histories, and relationships among plants can truly add to the enjoyment and appreciation of these amazing organisms. As in many parts of California, the natural areas around San Luis Obispo, and the plants living in them, have not escaped degradation or destruction. The need for conservation is more real now than ever, and we will only conserve what we know and appreciate. We are fortunate to live in an area where we are still surrounded by immense botanical diversity, novelty, and interest. There is much beauty right out our front doors, if we just remember to look.

Matt Ritter
San Luis Obispo, California, Fall 2005

Coastal Scrub & Chaparral

Coastal scrub occupies a narrow strip of land on dry hillsides, just off the coast, throughout most of California. Most dominant coastal scrub plants on the Central Coast are medium-size, soft-stemmed, drought-deciduous shrubs, some of which bloom with showy bright blue, red, and yellow insect pollinated flowers. The gentle rolling hills and mesas occupied by coastal scrub are prime California oceanfront real estate. For this reason, the plants and animals unique to this habitat are threatened by housing and commercial development and have disappeared completely from many areas of Southern California.

Chaparral covers almost one-tenth of our state, mostly in dry, interior, lower elevation areas. It is a thicket of impenetrable, large shrubs with woody branches and small, stiff, leathery leaves. Both chaparral and coastal scrub habitats are extremely prone to fires during the dry months. However, many of the shrubs have adaptations allowing them to quickly regenerate after a fire, such as resprouting from dormant buds on underground root crowns and seed germination in the presence of high heat or chemicals from burned wood. The following pages contain descriptions of the species that dominate the coastal scrub and chaparral plant communities in the San Luis Obispo area.

Yarrow
Achillea millefolium Carolus Linnaeus (1707-1778)
Asteraceae *(ah-KILL-lee-ah mil-lih-FOE-lee-um)*

Yarrow, found natively in many areas of the globe, is an aromatic perennial herb that grows in western portions of California from the Oregon border south to Santa Barbara, with different varieties being more common in certain habitats. Its flat top, umbrella-like clusters of tiny white flowers appear in the late spring through early summer. The renowned eighteenth-century Swedish botanist Carolus Linnaeus named this species *millefolium*, Latin for a thousand leaves, due to its tiny fern-like, divided leaves. The Greek warrior Achilles, for whom this genus was named, was said to use this plant to stop the bleeding of battlefield wounds. Yarrow, which was sometimes used in a decoction to help menstrual pains, supposedly has many medicinal properties including being antiseptic, anti-inflammatory, antispasmodic, analgesic, astringent, and anesthetic. The difference between food, medicine, and poison is often a matter of dosage, and yarrow should be used carefully as it contains a potent neurotoxin called thujone (see the California sagebrush description on page six for more about thujone).

Manzanitas
***Arctostaphylos* spp.** Michel Adanson (1727-1806)
Ericaceae *(ark-toe-STAH-fil-lus)*

The manzanitas are an extremely diverse group of shrubs and small trees in California, with fifty-eight of the sixty existing species growing in the state. They are often a major component of chaparral, and a number of species are capable of quickly resprouting after fire, while others are completely destroyed and only regenerate from seed. The easy-to-recognize manzanita genus, *Arctostaphylos*, has approximately fifteen species in San Luis Obispo County, all of which look quite similar and often hybridize, making them difficult to distinguish. Some are highly localized to certain areas, such as the Morro Bay manzanita (*A. morroensis*, pictured bottom right), which is found only on the south side of Morro Bay. Most species have satiny, smooth, reddish-brown bark covering crooked stems and leathery, evergreen, vertically oriented leaves, which help prevent desiccation. Clusters of urn-shaped pink or white flowers develop into brownish red fruit, which resemble small apples (manzanita means little apple in Spanish). *Arctostaphylos* is the combination of the Greek words *arktos*, a bear, and *staphyle*, a bunch of grapes, a name referring to the fact that bears and other wildlife relish the abundant fruit. These fruit were also pulverized fresh to make a drink, ground into flour that was baked into cakes, and eaten dried by aboriginal Californians.

Coastal Scrub and Chaparral

Chamise
Adenostoma fasciculatum
William Hooker (1785-1865) & George Arnott (1799-1868)
Rosaceae *(ad-dih-NAW-steh-muh fah-sih-kew-LAY-tum)*

The word chaparral means "where scrub oaks grow," a derivation of the Spanish term for scrub oaks, *chaparro*. However, a more accurate name would probably be "where chamise grows" in that it is one of the most characteristic species of the chaparral. Often its presence defines a shrubland as chaparral. This species is abundant on dry ridges and steep slopes away from the immediate coast, where dense, pure stands blanket foothills with uniform, dark green vegetation. These stands are extremely flammable during the summer months and the resinous plants burn regularly. Following a fire, chamise seeds germinate prolifically, and surviving, underground burls of mature plants resprout vigorously. Chamise grows as a large evergreen shrub with rigid, woody branches and needle-like leaves arranged in small bundles (fascicles), from which it gets its species name. Small cream-colored flowers cluster on shoot tips from April into early summer. Aboriginal Californians used chamise hardwood shoots for clam gathering sticks, arrow shafts, digging and reaming tools, as well as firewood, for which the Spanish term is *chamiso*. *Adenostoma* is Greek for glandular mouth, in reference to minute glands on the mouth of certain floral organs.

California Sagebrush
Artemisia californica Christian Lessing (1809-1862)
Asteraceae (ar-teh-MEE-zsa kal-lih-FOR-ni-kuh)

California sagebrush, which ranges from Marin to Baja California, is an extremely common shrub in the coastal scrub plant community. It has slender, flexible stems and thread-like, very fragrant leaves, which are covered with so many fine hairs that they appear gray in color. Like a number of coastal scrub plants, California sagebrush is drought-deciduous, meaning it loses its leaves during the dry summer months. Although it is called sagebrush due to its scent, it is not a true sage (*Salvia* spp.), but a member of the sunflower family. Numerous plants in the genus *Artemisia* have medicinal value and yield stimulants as well as drugs for curing wounds and intestinal worms. The active ingredient in the French drink absinthe, a chemical named thujone, comes from a related species called wormwood (*Artemisia absinthium*). The stoic but volatile Dutch impressionist painter Van Gogh was under the influence of this drink when, on Christmas Eve 1888, he sliced off the lobe his left ear and sent it to a local prostitute. This genus is named in honor of Artemis, the Greek goddess of chastity.

Locoweed

Astragalus spp. Carolus Linnaeus (1707-1778)
Fabaceae *(as-TRAG-guh-lus)*

Locoweeds are members of the pea and bean family and belong to one of the largest known genera, with about 2,000 species worldwide. There are nearly 100 California species in this incredibly diverse genus, a number of which are rare and localized to a specific habitat. These species can either be annuals or perennials, both having hairy, pinnately compound feather-like leaves divided into many small leaflets arising from a center axis. Clusters of pea-like flowers appear in spring and are followed in the summer by dry pods. In some local species, seeds can be heard rattling inside the inflated pods (some go by the common name rattleweed). Certain locoweeds are toxic, either because they extract and accumulate selenium from the soil or make caustic alkaloids. The Spanish word for crazy, *loco*, was bestowed on these poisonous plants after settlers observed that domestic animals that ingested them exhibited strange behaviors. *Astragalus* comes from a Greek word that means both anklebone and dice. The latter is more appropriate and was probably used when naming the genus to describe the rattling of the squarish seeds inside the dried pod. The two species of locoweed pictured here are *A. curtipes* (background and top left), and *A. nuttalii* (right).

Coyote Brush

Baccharis pilularis Augustin Pyramus de Candolle (1778-1841)
Asteraceae (BACK-kuh-riss pill-yew-LAR-iss)

Coyote brush, a dioecious member of the sunflower family and one of the most common shrubs in California, can be found as a major component of coastal scrub throughout the western California coastal ranges. Due to its ability to quickly colonize disturbed areas, this leggy, evergreen shrub is extremely prevalent along freeways and roadsides. Its tiny egg-shaped leaves are often sticky and its abundant off-white floral heads emerge in the fall. A decoction of coyote bush leaves was used by the Chumash to treat poison oak rash. Carolus Linnaeus, the eminent eighteenth-century botanist, who introduced the Latin binomial system of naming different species, named *Baccharis* in honor of Bacchus the Greek god of wine. Augustin Pyramus de Candolle, an early nineteenth-century French botanist, named this species *pilularis*, meaning little balls, probably referring to this shrub's small spherical floral buds.

Wild Morning Glory
Calystegia macrostegia (Greene) Richard Brummitt (1937-)
Convolvulaceae *(kal-lih-STEE-gee-uh mack-roe-STEE-gee-uh)*

The large morning glory family (Convolvulaceae), with over 1,600 species, is mostly tropical and includes economically important members like the sweet potato. Wild morning glory, a California native and relative of the pernicious bindweeds, is a scrambling, herbaceous vine with long shoots emerging from a woody base. This species grows in dry, often rocky, coastal areas, particularly in coastal scrub plant communities from San Francisco south to Mexico. Its slender stems are adorned with light green, triangular, lobed, arrow-shaped leaves, sometimes covered with fine hair. The flowers, which begin to appear in late winter, have five fused petals that form a white, funnel-shaped corolla with light pink to purple stripes. This species can be highly variable in appearance and a number of subspecies exist. *Calystegia* comes from the Greek words *kalyx* and *stegon*, meaning a calyx cover, describing the leaf-like structures that emanate from the stem of some species just below the flower and cover the calyx. *Macrostegia* means large covering.

Buck Brush
Ceanothus cuneatus (Hooker) Thomas Nuttall (1786-1859)
Rhamnaceae *(see-ah-NOE-thuss kew-NEE-aye-tus)*

There are over forty species of *Ceanothus* ranging from Canada to Guatemala, most of which are native to California. They are common species in the fire-prone chaparral and the seed coats of some must come in contact with smoke before germinating. Buck brush, a stiff shrub with small, evergreen, opposite, sessile leaves, is the most common *Ceanothus* in San Luis Obispo County. It lives throughout California, southern Oregon, and Northern Baja California, sometimes forming large impenetrable chaparral thickets. The showy, late winter flowers range from white to blue to lilac and give off a distinctively musky scent. Aboriginal Californians ate the seeds and brewed a tea from the leaves to treat coughs, fevers, and colds. The blossoms make a soapy lather when rubbed on the skin, and the roots produce a red dye. *Ceanothus* is the Greek word for spiny plant and *cuneatus* means wedge shaped, describing the leaves.

Mountain Mahogany
Cercocarpus betuloides John Torrey (1796-1873) & Asa Gray (1810-1888)
Rosaceae *(ser-koe-KAR-pus bet-choo-LOY-deez)*

Mountain mahogany grows as a large evergreen shrub or small tree in chaparral and dry woodlands throughout most of California and into Oregon and Baja California. Its leaves, which have toothed margins only on the upper half, are dark green, sticky, and hairless on top with woolly undersides. These leaves have the superficial appearance of birch (*Betula* spp.) leaves, leading to the species name, which means birch-like. The off-white, fragrant, petal-less flowers appear in clusters of two or three from March to May. This species' common name is derived from the hardness of the wood, which aboriginal Californians found many uses for, including digging sticks, fish spears, smoking pipes, clubs, and arrow tips. *Cercocarpus* is Greek for tailed fruit, describing the long, twisted floral style, which is covered with silver hairs (upper right photo). During the summer, when mountain mahogany is heavily visited by foraging wildlife, these feathery styles, which remain attached to the single seeded fruit at maturity, help in seed dispersal by attaching to animal fur.

Live Forever or Dudleya
Dudleya **spp.** Nathaniel Britton (1859-1934) and Joseph Rose (1862-1928)
Crassulaceae *(DUD-lee-uh)*

There are a number of distinct species of *Dudleya* in and around San Luis Obispo as well as several sub-species and hybrids between species. This fact has made their botanical classification difficult. Most have thick, water-storing, succulent leaves allowing them to live on harsh, south-facing slopes, often crouching in crevices on open rock. The small, usually red, orange, and yellow flowers are borne on the end of a shoot that appears in the spring and early summer. The plump leaves, which in some species are covered with a chalky, waxy powder, grow near the ground in symmetrical, basal rosettes. One species shown here (*D. abramsii*, top right) is a rare species that grows only on the bluish-grey serpentine soils in the foothills around San Luis Obispo. The other species pictured here are the chalk dudleya (*D. pulverulenta*, top left), and the lance-leafed dudleya (*D. lanceolata*, bottom right), and sea lettuce (*D. caespitosa*, bottom left and opposing page). The genus is named after William Russell Dudley, one of the founders of the Sierra Club and head of the botany department at Stanford University until his death in 1911.

California Buckwheat
Eriogonum fasciculatum George Bentham (1800-1884)
Polygonaceae (*air-ee-AH-goh-num fas-ick-yew-LAY-tum*)

California buckwheat is the most widespread member of the largest dicot genus (over 250 species) in California. Members of the four varieties of this species, which are distinguished by leaf characteristics and native range, occur on dry slopes, in chaparral, coastal scrub, desert areas, canyons, and washes throughout California, the Southwestern U.S., and Northwestern Mexico. The species name of this low-growing, slightly woody shrub alludes to the leaves, which are clustered in fascicles on the stem. These leathery, evergreen leaves have edges rolled under hiding woolly undersides. The small, light brown or pink, funnel-shaped flowers appear in terminal clusters in spring and through the summer in some areas. Although no part of this plant is especially edible, the flowers are an important food source for butterflies and honeybees. Decoctions of the roots and dried floral heads were used by aboriginal Californians to treat a variety of ailments. Both rhubarb (*Rheum*) and buckwheat flour (*Fagopyrum esculentum*) come from other members of the same plant family. The genus name, *Eriogonum*, means woolly knees, describing the swollen, hairy nodes of some of the first species ever described.

Coastal Scrub and Chaparral

Golden Yarrow
Eriophyllum confertiflorum (DC) Asa Gray (1810-1888)
Asteraceae *(air-ree-oe-FIL-um con-fer-tih-FLOOR-um)*

Golden yarrow, a member of the sunflower family, is a small, grayish-green, perennial subshrub ranging from Mendocino County south to Baja California. On the Central Coast, this species is common in many habitats including stabilized dunes, oak woodlands, coastal scrub, and chaparral. Its deeply lobed leaves are covered with minute, matted hairs, giving the plant its woolly, gray appearance. The flowers appear in bright yellow, flat top clusters of sunflower-like heads from late winter until early summer. The common name describes the resemblance of its small floral heads to those of true yarrow (*Achillea millefolium*). *Eriophyllum* means woolly leaf and *confertiflorum* means crowded flowers.

Lizard Tail or Seaside Woolly Sunflower
Eriophyllum staechadifolium Mariano Lagasca y Segura (1776-1839)
Asteraceae *(air-ree-oe-FIL-um stay-kad-dih-FOE-lee-um)*

Lizard tail is a small, highly-branched, perennial shrub. It is a component of the coastal scrub plant community, especially in Northern California. It reaches its southern limit in Santa Barbara County and ranges northward along the coast to Coos County, Oregon. In coastal areas of San Luis Obispo, it can be found on the edges of beaches, stabilized dunes, and coastal bluffs, never extending more than a few miles inland. The grayish-green, fragrant leaves are deeply divided and hairless on top with extremely woolly undersides, and rolled under edges. Starting in April and continuing throughout the summer, lizard tail produces bright yellow sunflower inflorescences composed of many minute ray and disk flowers (see the tidy tips [*Layia platyglossa*] description on page 100). The Northern California coastal Miwok Indian Tribe ground lizard tail seeds into a mush, called pinole, and placed leaves on the skin to relieve aches and pains. This is an easy-to-grow plant and is popular in coastal gardens where vibrant golden blooms contrast well with the frosty, gray foliage. *Eriophyllum* is from the Greek words *erion*, meaning wool, and *phyllon*, a leaf. *Staechadifolium* means having leaves like *Stoechas*, from an ancient Greek word for lavender and a genus name that no longer exists.

Coastal Scrub and Chaparral

18 *Coastal Scrub and Chaparral*

Our Lord's Candle or Chaparral Yucca
Hesperoyucca whipplei (Engelmann) John Baker (1834-1920)
Agavaceae *(hess-pare-oe-YUK-ah wipp-LEE-eye)*

Our lord's candle is a trunkless shrub with long bayonet-like leaves arising from a ground level rosette. This species blooms from March to July in chaparral and coastal scrub, throughout Southern California and Northern Baja California. A magnificent flower stalk, often more than ten feet tall, adorned with fragrant cream-colored flowers emerges from the rosette of leaves. After flowering and fruiting occur, the rosette and shoot die while only basal offsets live on. This plant has a truly remarkable symbiotic relationship with the yucca moth (*Tegeticula maculata*), which both species depend on for their reproduction. Each spring a pregnant yucca moth collects specially formed pollen into waxy balls before laying her eggs into the ovary of the flower. She then proceeds to push the pollen balls into the stigma at the top of the ovary, ensuring a good pollination and the development of seeds, some of which will serve to feed her offspring. Aboriginal Californians used the seeds for flour and leaf fibers for cordage. Once considered a member of the genus *Yucca*, a name derived from the Carib Indian word for cassava, this species was moved to the genus *Hesperoyucca*, meaning western yucca, because DNA evidence and numerous characteristics separate it from other yuccas. *Whipplei* commemorates A.W. Whipple, an early Californian railroad surveyor.

Saw-Toothed Goldenbush
Hazardia squarrosa (Hook. & Arn.) Edward Greene (1843-1915)
Asteraceae *(ha-ZAR-dee-uh skwar-ROE-suh)*

Saw-toothed goldenbush is a small, multi-stemmed shrub with a woody base and herbaceous stems, often called a subshrub. It is a component of coastal scrub and chaparral and has a native range mostly near the coast from Monterey County south into Baja California. Saw-toothed goldenbush earns it common name in part from its small, oval, clasping leaves, which are sharply toothed along the margins with spiny tips. The second half of the common name comes from the golden floral heads, which are clusters of many tightly packed disk flowers. *Hazardia* is one of a number of genera that was formerly placed in the large genus *Haplopappus*, which is now limited to South American plants. The name honors Barclay Hazard, a nineteenth-century amateur Santa Barbara botanist who inspired Edward Greene, whom named the species, to visit the Channel Islands. Edward Greene, who described and named many California plants, was the first plant taxonomist at U.C. Berkeley and predecessor to Willis Linn Jepson, possibly California's most influential plant biologist. *Squarrosa* refers to the leaves, meaning scaly and rough.

Toyon
Heteromeles arbutifolia (Lindley) Johann Roemer (1763-1819)
Rosaceae *(het-ter-RAW-mih-leez ar-BEW-tih-foe-lee-ah)*

Toyon is a large shrub or small tree, which grows in chaparral, oak woodlands, and evergreen forests throughout California and Northern Baja California. It is one of the most attractive, well-known, and common native plants in California. Through the summer, toyon develops clusters of white flowers followed in the fall by pea-sized bright red fruit. If not picked or eaten by birds, these fruit will stay on the plant until early winter. The genus name, *Heteromeles*, means different apple, a name used to describe the fruit (called a pome), which is botanically similar in structure to apples and pears. At times, entire limbs bearing these beautiful fruit were gathered from the wild and sold commercially for floral trimming around Christmas time (another common name is Christmas berry). However, this practice was outlawed in the state many years ago. When early settlers in the hills above Los Angeles saw this shrub growing in the surrounding chaparral, with its dark, evergreen, toothed leaves and bright red fruit, they were reminded of the east coast hollies, and the name Hollywood was chosen for the area. The species name, *arbutifolia*, refers to the similarity between the leaves of toyon and leaves of members of the genus *Arbutus*, which includes the Pacific madrone tree (*Arbutus menziesii*).

Climbing Penstemon or Heart-Leaved Penstemon
Keckiella cordifolia (Bentham) Richard Straw (1926-)
Scrophulariaceae *(keck-ee-EL-uh kor-dih-FOE-lee-uh)*

There are seven species of *Keckiella* and all are found in California. In a 1960s revision of the genus *Penstemon*, Dr. Dick Straw of Cal State Los Angeles separated the woody, shrubby penstemons into the genus *Keckiella*, a name that honors David Keck, a twentieth-century taxonomist. Originally the name given to the new genus was *Keckia*, until it was discovered in 1967 that this name had been previously used for a genus of extinct fossil algae. Because the same genus name can not be used to describe two unrelated species, *Keckiella* was created. Climbing penstemon grows in oak woodlands, coastal scrub, and chaparral throughout Southern California and Baja California, eventually reaching its northern limits around San Luis Obispo. This is a native, perennial, drought-deciduous, woody, vining shrub with spreading, wand-like, often hanging stems. Attached to these stems are dark green, heart-shaped leaves (*cordifolia* is Latin for heart-shaped foliage) with toothed margins, arranged oppositely, two at each node. This species is rather inconspicuous until late spring and summer when it produces clusters of bright reddish-orange flowers with a long, two-lipped floral tube.

Honeysuckle
***Lonicera* spp.** Carolus Linnaeus (1707-1778)
Caprifoliaceae *(loan-ISS-er-uh)*

Honeysuckles are erect shrubs or slender, sprawling, woody vines that climb over other shrubs and trees. They have rounded leaves that emerge from the stem in opposite pairs. In some species the uppermost pair of leaves are fused around the stem and in others they are free. The tubular flowers, which are clustered in terminal spikes, appear in late spring or early summer, and are followed by small berries. The Yuki and Shoshone tribes crushed leaves and pounded roots of the chaparral honeysuckle to treat sores and swelling. There are four native honeysuckles that one may encounter in the Central Coast area. Twinberry (*L. involucrata* var. *ledebourii* with paired flowers), the chaparral honeysuckle (*L. interrupta*, with yellow flowers), California honeysuckle (*L. hispidula*, with pink flowers, pictured here), and southern honeysuckle (*L. subspicata*), which has upper leaf pairs that are not fused. The horticulturally grown Japanese honeysuckle (*L. japonica*) sometimes escapes from cultivation especially in urban and riparian areas. *Lonicera* was named in honor of Adam Lonitzer, a sixteenth-century German physician who, like most physicians in his day, was also an herbalist and botanist.

Deer Weed
Lotus scoparius (Nuttall) Alice Ottley (1882-1971)
Fabaceae *(LOW-tus skoe-PAIR-ee-us)*

Deer weed, also known as California broom, is a very common, perennial subshrub with spreading, wand-like branches and a slightly woody base. It is distributed throughout the coastal foothills, on stabilized dunes, in coastal scrub, and is often one of the first shrubs to colonize a disturbed or recently burned area. The compound leaves, made up of three small leaflets, wither and fall off in the summer giving this bush the appearance of a broom. Pea-like, bright yellow flowers appear whorled around the stems in early spring and change in color to orange and red once they become pollinated. The roots of this species, as in many other legumes, enjoy a symbiotic relationship with nitrogen-fixing bacteria that convert nitrogen gas in the atmosphere to usable forms for the plant. This relationship could be partly responsible for deer weed's ability to quickly exploit burnt and disturbed areas. *Lotus* is a classical Greek name originally applied to many different unrelated plants. *Scoparius* means broomlike in Latin, alluding to this plant's appearance when drought-deciduous.

California Man-Root
Marah fabaceus (Naudin) Edward Greene (1843-1915)
Cucurbitaceae *(MAR-uh faw-BAY-shus)*

California man-root, sometimes called wild cucumber, belongs to the same family as melons, gourds, squash, and pumpkins. It is a native, perennial, herbaceous vine with large palmately lobed leaves and tendrils that help it climb over coastal scrub shrubs or amongst trees in riparian areas throughout California. The name man-root comes from its huge underground root, which can be as large as a buried human corpse. The star-shaped, cream-colored flowers, which are separated either male or female, on the same plant (monoecious), appear from January to April. Once fertilized, the ovary within a female flower turns into a bright green, prickly, spherical fruit the size of a golf ball or larger. Aboriginal Californians used the beautiful marble-sized seeds inside the fruit to make necklaces. A decoction of the vine has been said to cure venereal diseases, and oil from the seeds supposedly slows balding. *Marah* means bitter in Latin, alluding to the fact that no part of this plant is palatable. The large seeds are *fabaceus*, meaning bean-like.

Sticky Monkey Flower
Mimulus aurantiacus William Curtis (1746-1799)
Phrymaceae *(MIM-yew-lus our-ran-TIE-ah-kuss)*

Sticky monkey flower is an evergreen subshrub found in coastal scrub, chaparral, and wooded areas throughout much of California. The leaves are sticky to the touch and rolled under at the edges. The yellow to apricot-colored, tubular flowers have a white stigma with sensitive lips that close when touched or pollinated. This species has many different regional forms with various floral colors and shapes; so plants of the same species can look very dissimilar in San Diego, Santa Barbara, San Luis Obispo, or Santa Cruz. Aboriginal Californians ate the young leaves as salad greens and used crushed leaves and stems to sooth wounds and burns. *Mimulus* is either derived from the Latin word *mimus*, a mimic, or the Greek word, *mimo*, an ape, referring to the similarity between the flowers and the face of a mime or a monkey. *Aurantiacus*, meaning golden, refers to the floral color.

California Peony
Paeonia californica John Torrey (1796-1873) & Asa Gray (1810-1888)
Paeoniaceae *(pay-OH-nee-ah kal-lih-FOR-ni-kuh)*

The California peony is a handsome, perennial herb, usually less than two feet tall, growing in the chaparral and coastal scrub understory from Monterey County south into Baja California. Its fleshy, deeply divided leaves are bluish-green on top and pale on the underside. The large, solitary, spherical flowers droop heavily downward, almost seeming to burden their stems, never completely opening. Their striking petals are blackish maroon with red or pink margins. This species flowers early in the year and the blooms, usually finished by mid-March, are followed by a number of cone-shaped, pointed seedpods. Early Californians found the peony root helpful in treating sore throats, stomach ailments, and colds when eaten raw, dried and ground into a powder, or taken as a tea. *Paeonia* is derived from Paeon, physician to the Greek Gods, alluding to the medicinal values of many of the thirty members of this mostly Eurasian genus.

Holly-Leafed Cherry or Islay
Prunus ilicifolia ssp. *ilicifolia* (Nuttall) Wilhelm Walpers (1816-1853)
Rosaceae (PREW-nus il-lih-sih-FOE-lee-uh)

The genus *Prunus* has over 200 species, including important Asian members like almonds, apricots, peaches, plums, and cherries. There are only eight species native to the state, and holly-leafed cherry, the only species with evergreen leaves, is the most common in Central and Southern California. It grows in chaparral and woodlands from Napa and Sonoma counties south to Baja California. This large, woody shrub, has dark, leathery leaves with spiny edges reminiscent of holly leaves. Small white flowers in a long cluster emerge in the spring and are followed by dark purple stone fruits (drupes) similar to cherries, but with larger seeds and less pulp. Nevertheless, this fruit was an important food source for a number of aboriginal Californian tribes who ate them fresh or crushed and fermented into an alcoholic beverage. Teas made from the bark and leaves were also used to treat colds and headaches. Fruit pits were cracked and the seeds inside were crushed, processed in a number of ways to remove their bitter and toxic hydrocyanic acid, and then eaten as a mush. This species has been grown horticulturally for many years and will make a beautiful hedge or small tree depending on how it is pruned. *Prunus* is the Latin name for the plum, and *ilicifolia* means holly-like leaves.

Everlastings or Cudweeds
Pseudognaphalium spp. Carolus Linnaeus (1707-1778)
Asteraceae *(sue-doe-naf-FAL-ee-um)*

There are approximately 100 species in the cudweed genus, *Pseudognaphalium* (previously *Gnaphalium*). They are distributed throughout the world with about ten species, both native and introduced, occurring in California, most of which occur in the San Luis Obispo area. Like other members of the sunflower family, the flowers of these short-lived herbs are produced in compact heads. However, in this genus, the small heads are surrounded by more conspicuous, papery, white, light brown, or rose-tinted phyllaries. Everlastings are found in many different habitats and plant communities including open areas of coastal scrub and chaparral, dry wooded hills, grasslands, and on stabilized dunes and beach bluffs near the coast. The sessile leaves, which are sticky and strongly scented in some species, are arranged alternately and have smooth margins. The small, often egg-shaped floral heads are produced in branched clusters throughout the spring and summer. The species pictured here are (clockwise from top left) California everlasting (P. *californicum*), which is the most common everlasting on the Central Coast, Jersey cudweed (P. *luteo-album*), a weedy species, pink cudweed (P. *ramosissimum*), and felt-leaved everlasting (P. *beneolens*). *Psuedognaphalium* is derived from the Greek word *gnaphalon*, meaning a lock of wool, referring to the woolliness of the leaves of many species.

California Wild Rose

Rosa californica Adelbert von Chamisso (1781-1838) & Diederich Franz Leonhard von Schlechtendal (1794-1866)

Rosaceae *(ROE-suh kal-lih-FOR-ni-kuh)*

The rose family contains many economically important plants including strawberries, blackberries, apples, almonds, and peaches. In addition to these food crops, there are also the horticulturally important roses in the genus *Rosa*. The California wild rose, although not of great economic value, is a beautiful, perennial shrub with a native range throughout western California including Northern Baja California and southern Oregon, frequently growing in riparian thickets. It has stout, slightly curved prickles and leaves that are pinnately divided into five or seven toothed leaflets. The large, mildly fragrant flowers, which appear in late spring, have five pink petals surrounding many yellow stamens. Once pollinated, these flowers develop into a strange kind of fruit called a hip. An individual rose hip is actually fleshy tissue derived from fused bases of the floral organs. This tissue surrounds many tiny, one-seeded fruits, called achenes (pronounced ackeens). Wild rose hips, which supposedly have high vitamin content, can be eaten or dried for tea. *Rosa* is the ancient Latin name for a rose.

Chia

Salvia columbariae George Bentham (1800-1884)

Lamiaceae *(SAL-vee-ah koe-lum-BAR-aye)*

There are approximately 900 species in the mostly tropical genus *Salvia*, with eighteen native to California. Most of the California *Salvias* (sages) are perennial shrubs with fragrant leaves attached oppositely, two at a time, to square stems. Chia, one of only two annual California sages, is usually less than two feet tall with rough, dull green, deeply lobed leaves mostly growing at ground level. It grows away from the immediate coast on dry hillsides in chaparral, coastal scrub, and disturbed areas in California from Mendocino south into Mexico and throughout the southwest. In mid-spring, a flowering stem emerges from the basal leaves bearing spiny, whorled tiers of flowers with dark blue petals fused into trumpet shaped tubes with lobed tips. Below the flowers are clusters of minute, sharply pointed, wine-colored bract leaves. The tiny, nutritious seeds, which can be harvested by shaking a dried floral head over an open palm, were a staple food for many aboriginal Californian tribes. They were eaten raw, roasted, ground into meal, or added to water to make a refreshing drink. The name *Salvia* is derived from the Latin word *salvus*, meaning safe or well, referring to the medicinal value of members of this genus. George Bentham, a nineteenth-century British botanist, may have named this species *columbariae* due to similarities with the Mediterranean native *Scabiosa columbaria*, a horticultural plant in the teasel family (Dipsacaceae).

Black Sage
Salvia mellifera Edward Greene (1843-1915)
Lamiaceae (*SAL-vee-ah meh-LIH-fur-uh*)

Black sage is an important component of both coastal scrub and chaparral plant communities. This deciduous, drought-tolerant shrub produces pale blue, ball-like clusters of flowers in late spring, before losing its leaves during the long, dry California summer. It can be found from Central California south to Northern Baja California, sometimes in almost pure stands. A tea made from the aromatic leaves can be used as an antiseptic mouthwash and general disinfectant. The name *Salvia* is derived from the Latin word *salvus*, meaning safe or well, referring to the medicinal value of members of this genus. *Mellifera* means honey bearing, from the Latin words *mellis*, meaning honey, and *fera*, meaning bearing or carrying.

Blue Elderberry
Sambucus mexicana Jan Presl (1791-1849)
Adoxaceae *(sam-BEW-kus mex-ih-KAY-nuh)*

The blue elderberry is a native, winter-deciduous, multi-trunked shrub that can reach heights greater than twenty-five feet. It may be considered a small tree, except that it rarely has only a single trunk, a feature that many botanists use to help define the often-unclear distinction between shrubs and trees. It is distributed throughout western Canada, the United States, and Mexico, from where it derives its species name, *mexicana*. Blue elderberry is commonly found along streams and more sporadically in drier habitats such as coastal scrub and chaparral. Its compound, oppositely arranged leaves (two attached at one point on the stem) are divided into three to nine toothed leaflets. In late spring, this species produces large flat-topped clusters of many small cream-colored flowers. These are followed by bunches of small, dark blue, wax covered berries. The fruits, which are relished by birds, were also eaten by aboriginal Californians and are still used today for making pies, jellies, and wine. Cooked leaves and roots, as well as decoctions of the flowers, were used medicinally as external compresses to relieve soreness and inflammation. Aboriginal Californians also removed the spongy internal pith of new stems to make whistles, flutes, and other musical instruments. Coincidentally, Carolus Linnaeus named the genus *Sambucus* after a Greek musical instrument called the *sambuca*, made from the wood of a related species.

Nightshades
***Solanum* spp.** Carolus Linnaeus (1707-1778)
Solanaceae *(soe-LAY-num)*

The large and economically important nightshade family (Solanaceae) includes members such as tobacco, tomatoes, potatoes, eggplants, and peppers. There are two species in this family, both in the mostly tropical genus *Solanum*, which are common in Central Coast chaparral, coastal scrub, dunes, and disturbed areas. Purple nightshade (*S. xanti*) has saucer-shaped purple flowers with tiny white-encircled green dots in the center, while the white flowers of black nightshade (*S. douglasii*, the common name is for the fruit color) are smaller and star-shaped. The stamens of both species converge into a yellow cone around the pistil. The leaves and small black or purple tomato-like berries of these species are highly poisonous and should not be eaten. Juice from the berries was, however, used as tattooing ink by aboriginal Californians. *Solanum* means comforting in Latin, assigned to this genus because of the narcotic properties of some of its members. Both species are named after nineteenth-century botanists, Scottish plant collector David Douglas and Hungarian botanist John Xantus.

Other coastal scrub and chaparral plants. Clockwise from top left. Prickly phlox (*Linanthus californicus*), leather oak (*Quercus durata*), wart leaf ceanothus (*Ceanothus pappillosus*), bush poppy (*Dendromecon rigida*), flannelbush (*Fremontodendron californicum*), and chaparral pea (*Pickeringia montana*).

Oak Woodlands & Riparian

Oak woodlands are one of the most characteristic types of vegetation in California. Most oak woodlands in coastal San Luis Obispo County are dominated by the coast live oak (*Quercus agrifolia*). Coast live oak woodlands range from Sonoma County to northern Baja California, mostly occurring on north facing slopes and in canyons. The dense canopy of the coast live oak creates a very shady environment for understory species, most of which can tolerate low light. In many of these woodland areas the coast live oak is the only tree species. Some types of oak woodlands are among the state's most endangered plant communities due to livestock grazing of acorns and seedlings, land mismanagement, and urban development, and are a high priority for inclusion in future protected areas.

Riparian habitats are lush, green areas bordering waterways. Most of the large, broad-leaved trees and numerous shrubs that dominate this habitat depend on a permanent water supply and can be severely affected by floods and droughts. Riparian comes from the Latin word *ripa*, meaning a riverbank. Described on the following pages are the common species native to San Luis Obispo coast live oak woodland and riparian habitats.

Ferns

Ferns, including the natives shown here, are a large group of plants (over 11,000 species) that never produce seeds; instead they reproduce by spores. On the California Central Coast, native ferns grow from underground rhizomes in shady, moist environments, in the oak woodland understory, on north facing slopes, and in the cool cracks of rocky outcrops. All the terrestrial ferns native to San Luis Obispo have divided leaves, sometimes called fronds, which uncoil as they mature, exhibiting the appearance of a fiddlehead. At certain times, spores are present on the underside of these leaves in brown dot-like structures called sori (pronounced sore-eye), near the curled leaf edge, or scattered along the veins. Some of our most common ferns are (clockwise from top left) maiden-hair fern (*Adiantum jordanii*), coffee fern (*Pellaea andromedifolia*), California polypody fern (*Polypodium californicum*), and goldback fern (*Pentagramma triangularis*). Another locally abundant fern, bracken fern (*Pteridium aquilinum*, pictured on the opposing page), is one of the most common ferns in the world, with a range extending to most continents and many islands. The young fronds of many ferns can be eaten, but not in large quantities because some may contain toxins and cancer-causing chemicals, and bracken fern is poisonous.

Horsetails
***Equisetum* spp.** Carolus Linnaeus (1707-1778)
Equisetaceae *(eck-wih-SEE-tum)*

There are fifteen living species of horsetails, all in the genus *Equisetum*. These unusual plants are the remaining members of a once large and diverse group that has now mostly gone extinct. Their minute, papery, rudimentary leaves are fused into rings that whorl around nodes on green, vertically ridged, hollow stems. The abrasive cells on the outer stem surface are impregnated with glass crystals, which is why horsetails have been used historically for tasks such as polishing wood and silver and cleaning pots. This attribute is also the basis for their other common name, scouring rushes. Like the ferns, horsetails never make flowers or fruit. Instead they reproduce by spores produced at shoot tips in cone-like structures called strobili (one strobilus is pictured top left). Some horsetails have separate sterile and fertile shoots, such as the giant horsetail (*E. telmateia* ssp. *braunii*, background and right), while in others like the common scouring rush (*E. hyemale*, two photos on the left) all shoots produce strobili when mature. Horsetails grow in moist, sandy areas near marshes and waterways, and because they spread by rhizomes, they can create large clonal patches, sometimes becoming invasive. *Equisetum* is from the Latin words, *equus*, a horse, and *seta*, a bristle.

Mugwort

Artemisia douglasiana Joseph Gottlieb von Besser (1784-1842)
Asteraceae   (ar-teh-MEE-zsa duh-GLASS-ee-ay-nah)

Mugwort is an herbaceous perennial with erect stems that grow from an underground rhizome and large thin leaves smelling strongly of sage. It is common in canyons, drainages, riparian forests, and along stream courses throughout lower elevation areas of California. This species has very distinctive grayish-green leaves with three to five coarse lobes at the tip and fuzzy white undersides. Fresh juice from this plant has reportedly been effective at relieving the rash caused by exposure to poison oak (*Toxicodendron diversilobum*). Mugwort comes from the Anglo-Saxon terms *mug*, meaning midges or small flying insects, and *wort* meaning herb, alluding to the fact that long ago relatives of this plant were used to drive off fleas from bedding. This is one of many plants named after David Douglas, an ardent Scottish plant collector for the Royal Horticulture Society in England. He was the first person to collect plants in the San Luis Obispo area while traveling by foot from Monterey to Santa Barbara in 1831, during a two-year botanical exploration of the Pacific coast. He collected thousands of plants and seeds before attempting to return to Europe. During his voyage home, he was gored to death by a captive bull on a ranch near Hilo, on the Big Island of Hawaii, a few days after his 35th birthday.

Miner's Lettuce
Claytonia perfoliata Carl Ludwig von Willdenow (1765-1812)
Portulacaceae *(klay-TOE-nee-ah per-FOE-lee-ah-tah)*

Miner's lettuce is a common and widespread, small, annual plant found in vernally moist and shady areas throughout most lower elevation regions of California. It can be easily recognized by the fused pair of succulent leaves that create a round disk encircling the stem. Its small white or pink flowers sit above the leaf disks, which are edible and were eaten raw or boiled by aboriginal Californians and early miners. They make an excellent addition to salads. *Claytonia* is named after John Clayton, an eighteenth-century American botanist and son of the Attorney General for colonial Virginia. He collected many specimens that were sent to eminent European botanists. *Perfoliata* comes from the Latin terms *per*, meaning through, and *folia*, meaning leaf, used to describe the stem passing through the disk leaves.

Bedstraws
***Galium* spp.** Carolus Linnaeus (1707-1778)
Rubiaceae   *(GAL-ee-um)*

Bedstraws are a group of small, scrambling, or climbing plants that belong to the same large, primarily tropical plant family as coffee, gardenias, and *Cinchona*, the source of the malaria drug quinine. *Galium* is a large genus of over 300 species, with around fifty represented in California. They can be found in many habitats including coastal scrub, chaparral, and shady oak woodland understory. Many have square stems adorned with minute, downward pointing hook-like prickles that aid in attachment and climbing. Tiny flowers can be found in the axils of the small leaves, which are borne in whorls around the stems. The dried seeds and fruit of some species have been used as a coffee substitute. The common name comes from the long ago use of dried bedstraw for mattress stuffing and is believed to be the plant that filled the manger of baby Jesus. Three of the most common bedstraws native to the Central Coast are goose grass (*G. aparine*, picture on the left), climbing bedstraw (*G. porrigens*, pictured on the right), and prickly bedstraw (*G. andrewsii*). The genus name *Galium* is derived from the Greek word for milk, *gala*. *G. vernum*, a European bedstraw, was used to curdle milk during cheese making in England.

Common Monkey Flower
Mimulus guttatus Augustin Pyramus de Candolle (1778-1841)
Phrymaceae *(MIM-yew-lus guh-TAY-tus)*

The common monkey flower is a native, annual or perennial, non-woody plant that lives where soils are wet. This species is extremely variable in its size and can range from only a couple inches to over three feet tall. This wide variety in form may be genetically determined (subspecies have been recognized) or this may just be an extremely variable species that makes drastic growth responses to different levels of moisture and other environmental factors. The adaptability of the common monkey flower has probably helped it to spread throughout its large range in western North America, from northern Mexico to Alaska. On the Central Coast it can be found along streams, in wet fields, and any temporarily or permanently wet area where water seeps from the ground. Its broad, oval, irregularly toothed leaves emerge as opposite pairs on the stem. During common monkey flower's long flowering season, from early spring to fall, bright yellow flowers with two lips are produced on slender stalks. A close look reveals dots of red spotting the floral tube, a feature used in giving this species its name; *guttatus* is Latin for speckled. The calyx becomes swollen and brown as a capsular fruit, which is full of tiny seeds, develops inside (top left photo). *Mimulus* is either derived from the Latin word *mimus*, a mimic, or the Greek word, *mimo*, an ape, referring to the similarity between the flowers and the face of a mime or a monkey.

California Sycamore
Platanus racemosa Thomas Nuttall (1786-1859)
Platanaceae (*PLAT-tuh-nus ray-see-MOE-suh*)

San Luis Obispo's tallest native tree, the fast-growing California sycamore, can reach nearly 100 feet. This species is dispersed in canyons and near waterways in wetter, warmer parts of California and Baja California. The large velvety leaves are shed during the winter and spherical fruit clusters follow inconspicuous early spring flowers. The rigid older bark, which is incapable of expanding with the growing tree, flakes off in large sections, exposing younger grey patches. As bark is shed, the limbs take on their characteristic scaly, mottled look. The gnarled appearance of many of these trees is due to repeated infections by anthracnose fungus, which can completely defoliate trees during wet spring years. Few trees die, however, because a second crop of leaves is produced in the early summer. A related tree, which is common in urban areas, is the London plane tree (*Platanus x acerifolia*), a horticultural hybrid and likely the most common street tree in the temperate world. The name *Platanus* is derived from the Greek word for broad, *platys*, describing the large leaves of trees in this genus. *Racemosa*, which is derived from the Latin word *racemus*, a bunch of grapes, is used to describe the shape of this tree's floral clusters (inset photo).

Foothill or Gray Pine
Pinus sabiniana David Don (1800-1841)
Pinaceae *(PIE-nuss sah-BIN-ee-ah-nuh)*

There are nearly 100 species of pines with eighteen native to California. This is a very ecologically and economically important group of trees. Their long, slender, evergreen leaves, called needles, are formed in bundles of one to five depending on the species. The foothill pine, which grows natively only in California, can be distinguished from other pines by its multiple, leaning, crooked, black-barked trunks, irregular branches, and sparse foliage. Its grayish-green needles are bundled in threes, and the large seed cones are armed with flat, sharp, dangerous looking prickles. The foothill pine commonly grows with blue oak (*Quercus douglasii*) in foothill woodlands throughout much of California. It is locally abundant north of the Cuesta Grade and in the Santa Lucia Range. The large, nutritious seeds were an important food source for a number of aboriginal Californian tribes. Digger pine, a once widely-used common name, comes from a pejorative, colloquial pioneer term that groups all California Indian tribes together as "Digger" Indians, based on their practice of digging for foods such as roots and bulbs. This name is considered offensive, and its use is now best avoided. *Pinus* is the ancient Latin name for pine trees. David Don, a British botanist and librarian for the Linnean Natural History Society in London named this species in 1832 in honor of Joseph Sabine, a London lawyer, famed naturalist, and co-founder of the society.

48 | *Oak Woodland and Riparian*

Coast Live Oak
Quercus agrifolia Luis Neé (1734-1807)
Fagaceae *(KWER-kuss ag-rih-FOE-lee-uh)*

There are nineteen species of oaks native to California, ten species grow as trees and nine as shrubs. All but one occur only in southwest Oregon, western California, and northern Baja California (an area called the California Floristic Province). The coast live oak is an evergreen tree that reaches heights of twenty to seventy feet and can live for hundreds of years. It is the backbone of many Central Coast oak woodland communities, a coastal California icon, and the basis for city names like Thousand Oaks and Oakland. The Chumash Native American Tribe in San Luis Obispo and Santa Barbara counties supported large population densities on a diet largely based on coast live oak acorns. Spanish settlers associated this tree with fertile land, and the Franciscan missions almost perfectly match the coast live oak's native range. *Quercus* comes from the Celtic words *quer* and *cuez*, meaning fine tree. *Agrifolia* has no meaning in Greek or Latin, and is believed to have come from a mistaken transcription of the original name *aquafolia*, which means holly-like leaves in reference to the spiny leaf edges.

Blue Oak
Quercus douglasii
William Hooker (1785-1865) & George Arnott (1799-1868)
Fagaceae (*KWER-kuss duh-GLASS-ee-eye*)

The blue oak is one of California's most drought-tolerant and widespread oaks. This species occurs only in California where it endures and thrives on the hot, dry foothills surrounding the Central Valley. It is locally common from the eastern slopes of the Santa Lucia Range to the dry interior of San Luis Obispo County. A mature blue oak is a medium-sized tree with a compact canopy of deciduous leaves that have a dull bluish-green tint, especially in the late summer. The blue oak acorn is an egg-shaped nut that sits in a cup made of scaly bracts. The acorns, which contain a single seed full of stored nutrients, were a staple in the diets of many aboriginal Californian tribes. Unfortunately, the survival of this and a number of other California oak species is in jeopardy. In many areas of the state, aging or fallen oaks are not being replaced by young trees. Even though seasons of healthy acorn production and germination occur regularly, weeds, livestock grazing, fire suppression, urban development, and increased predation on acorns have all significantly impacted oak regeneration in the last 100 years. This is one of the many species named in honor of David Douglas, a Scotsman who collected California plants around 1830.

Valley Oak
Quercus lobata Luis Neé (1734-1807)
Fagaceae *(KWER-kuss low-BAH-tuh)*

The valley oak is the largest oak in California and possibly the largest species in North America. These beautiful trees only occur in California, growing mostly in the Central Valley, Coast Ranges, and Sierra Nevada foothills. Locally, valley oaks are found mostly in the northern and central portions of San Luis Obispo County. Growing on fertile bottomland soils, these oaks can reach great size, age, and beauty. Their massive trunks are covered with thick, deeply checkered bark and support broad, rounded canopies of craggy branches and dark green foliage. The deciduous leaves are deeply lobed, a distinguishing feature noted by the Spanish botanist Luis Neé who collected valley oak in 1791 and later gave it the descriptive name *lobata*. Early Spanish explorers called these majestic oaks *robles*, due to their similarity to the English oak (*Q. robur*). This name is also the origin of the town name Paso Robles. The fact that this species prefers rich, lowland, valley soils has made it the victim of extensive removal for agriculture over the last 150 years. Now only a few large, remnant individuals remain in agricultural areas that were once the home of vast valley oak stands.

California Coffeeberry
Rhamnus californica Johann von Eschscholtz (1793-1831)
Rhamnaceae *(RAM-nus kal-lih-FOR-ni-kuh)*

California coffeeberry is a medium-sized evergreen shrub found in woodlands, riparian areas, chaparral, and coastal scrub in western portions of California. It has reddish-brown, flexible stems and bright, evergreen, hairless leaves. This shrub's numerous, inconspicuous, nectar-filled flowers give rise to pea-sized drupes that progress through many different colors before settling on black. Aboriginal Californians stripped and dried the bark to make a laxative brew to help with the constipation that results from a heavy acorn diet. Early European settlers made an unsuccessful attempt to use the bitter inedible fruit as a coffee substitute. *Rhamnus* comes from the ancient Greek name for buckthorn, a member of the same plant family. *Californica*, the species epithet of many plants living on the Central Coast, is used to denote plants that are from California or were first collected here.

Redberry
Rhamnus crocea Thomas Nuttall (1786-1859)
Rhamnaceae *(RAM-nus KROE-see-uh)*

Redberry, an easily distinguished relative of the California coffeeberry, is a small, native, evergreen shrub rarely growing larger than four feet. It lives on coastal hillsides as a component of coastal scrub, as part of oak woodland understory, and further inland as part of the chaparral. It ranges from northern Baja California to Mendocino County in Northern California. Redberry's small, rounded, thick, toothed leaves are born on rigid, densely spreading, branches that are often tipped with a stout thorn. In spring, sparse clusters of inconspicuous, green, petal-less flowers are produced in the leaf axils. After being pollinated these flowers develop into little, red, inedible fruit similar to a miniature cherry with two stone-like pits on the inside. Aboriginal Californians used crushed roots and bark of redberry medicinally to treat intestinal ailments. *Rhamnus* comes from the ancient Greek name for a buckthorn and *crocea* means yellow, maybe referring to the color of the underside of the leaves.

Fuchsia-Flowered Gooseberry
Ribes speciosum Frederick Pursh (1774-1820)
Grossulariaceae *(RYE-bees spee-see-OH-sum)*

Fuchsia-flowered gooseberry is a bristly, evergreen shrub with leathery, dark green, irregularly toothed leaves. The beautiful, bright red, tubular flowers, which are pollinated by hummingbirds, hang from the stems. The stamens, which are twice as long as the rest of the flower, hang down with bright yellow tips. To ward off herbivores, there are three stout spines emanating from each node. This plant can be found growing in coastal scrub, chaparral, and woodlands of the coastal mountain ranges from the San Francisco Bay Area south to Baja California. The genus name is from the Arabic word *ribas*, meaning sour, referring to the taste of the grape-sized, bristly fruit. *Speciosum*, meaning showy, is a fitting description for this striking plant.

Wild Blackberry

Rubus ursinus Adelbert von Chamisso (1781-1838) & Leonhard von Schlechtendal (1794-1866)

Rosaceae *(ROO-bus UR-sih-nus)*

The wild blackberry is a native perennial bush with straggling and climbing stems that often form thickets. It grows in moist canyons, woodlands, and along creeks throughout non-desert areas of California and all the way north to British Columbia. Like poison oak (*Toxicodendron diversilobum*), it has compound leaves divided into three leaflets. However, the two can be easily distinguished by the fact that the stems of wild blackberry are covered with thin, straight, bristle-like prickles. The white, often unisexual flowers, which emerge in March or April, have five petals and numerous stamens or carpels depending on their sex. In the female flowers, the carpels and the central receptacle they surround become the fruit, which are actually aggregations of many tiny one-seeded plum-like fruits called drupelets. These tasty little morsels, which are apparently the wild ancestors of the cultivated loganberries and boysenberries, are definitely worth competing for with the birds and wildlife in midsummer. *Rubus* is the ancient Latin name for blackberries and their relatives, while *ursinus* means relating to bears, which also feast on the fruit.

Arroyo Willow
Salix lasiolepis George Bentham (1800-1884)
Salicaceae *(SAY-liks lay-zee-oh-LEE-piss)*

Of the over 400 species of willows worldwide, approximately fifty are native to California. The most common willow in the Central Coast area, which has a large range throughout the Western U.S., is the arroyo willow. These small, fast growing, deciduous shrubs or trees are often dominant in wet areas and along stream courses. The unremarkable flowers, which are clustered together in structures called catkins (pictured top left), appear in late winter before new leaves emerge. Seed forming female flowers and pollen producing male flowers develop on separate plants (dioecious). The leaves are bright green on top and dull, pale-green below. Aboriginal Californians had many uses for this tree's pliable shoots, such as basketry and shelter construction. Stem cuttings, which readily form roots in water, are now frequently planted for the stabilization and restoration of waterways. The bark, which was also used for making rope, contains salicylic acid, the active ingredient in aspirin. *Salix* is the ancient Latin name for a willow and *lasiolepis* means woolly scale, referring to the fuzzy scale-like leaves that protect the dormant buds or the fuzzy catkin bracts.

Hummingbird Sage
Salvia spathacea Edward Greene (1843-1915)
Lamiaceae   (SAL-vee-ah spath-ah-SEE-ah)

Hummingbird sage is a beautiful, low-growing herb with large, oblong, puckered leaves, which are sparsely hairy on top and densely hairy on the underside. It can commonly be found growing in sheltered coastal woodland areas, ranging from Sonoma County south to Orange County. Its spike-like inflorescence, which is crowded with brilliant crimson flowers, appears in mid-spring. Hummingbirds seek out the copious amounts of nectar at the base of the trumpet-shaped flowers. Aboriginal Californians made a decongestant tea from the leaves. *Salvia* is derived from the Latin word *salvus* meaning safe or well, referring to the medicinal value of members of this genus. A related species from Mexico called diviner's sage (*S. divinorum*) is used in traditional shaman healing rituals as a hallucinogen. After eating a paste made from leaves, the shaman lies in darkness and silence with the patient, and if the visions are strong enough, the healer can find the cause of an illness. *Spathacea* comes from the Latin word *spatha*, a broad flat blade, probably alluding to the prominent, purple-bronze bracts that sit below the flowers.

Resurrection Plant
Selaginella bigelovii John Underwood (1770-1834)
Selaginellaceae (seh-LAH-jih-nell-ah big-LOE-vee-eye)

Resurrection plant, or spike moss, is a strange and primitive plant that generally resembles moss, although it is not a moss at all. Spike mosses have vascular tissue (xylem and phloem) for transporting water and sugar, where as true mosses do not. However, like true mosses, spike mosses never form seeds or fruits. Instead they reproduce by spores that are born in small cones at the shoot tips, looking like little spikes. There are over a thousand spike moss species in the world, but this is the only one native to the Central Coast. This species is commonly found forming carpets on serpentine outcrops, sandstone boulders, exposed rock crevices, and other rocky places throughout much of western California. If you were to see dried up, dormant resurrection plant for the first time in summer or fall, you would think it was dead. However, after the first rains, it quickly springs back to life, turning bright green. *Selaginella* is derived from the name of a distantly related plant *Lycopodium selago*, while the species name commemorates Dr. John Bigelow, a surgeon and botanist from Ohio who collected California plants during the transcontinental railroad survey in the 1850s.

California Hedge Nettle
Stachys bullata George Bentham (1800-1884)
Lamiaceae *(STAY-kees BULL-ah-tah)*

California hedge nettle, like many members of the mint family, has stems that are square in cross section and oppositely arranged leaves, two attached at every node. This hairy, low-growing species, rarely taller than a couple of feet, grows in the shady understory of oak woodlands and coastal scrub from Monterey County south to Baja California. The lavender flowers, which usually appear in March, are whorled in tiers in a spike-shaped inflorescences at the top of the plant. This plant's misleading common name comes from the stiff white hairs that cover the flexible leaf blades, making it reminiscent of stinging nettle (*Urtica dioica*). However, unlike stinging nettle, nothing but enjoyment can come from touching these soft, pleasant smelling leaves. *Stachys* is Greek for an ear of corn, possibly used to describe the shape of the spike-like inflorescence of another species of this genus. *Bullata* comes from the Latin word *bullate*, meaning puckered or blistered, alluding to the leaf texture.

Snowberry
***Symphoricarpos* spp.** Thomas Nuttall (1786-1859)
Caprifoliaceae   *(sim-for-rih-KAR-poes)*

There are two snowberries on the Central Coast that are distinguished by their size. Both are deciduous shrubs, which live near streams and in the understory of oak woodlands, with small, thin, oppositely arranged leaves. Common snowberry (*S. albus* var. *laevigatus*), an erect shrub, is the larger of the two species, usually growing between two and six feet tall. Creeping snowberry (*S. mollis*, pictured here) is a smaller, sprawling plant, usually less than two feet tall. In the spring and early summer, both species make clusters of tiny, pinkish red flowers. The flowers are followed by bright white, pea-sized fruit, which can be quite striking in the fall after all the leaves have fallen. The mealy fruits are inedible and possibly toxic, which may explain their persistence on leafless stems, even in the face of marauding wildlife. *Symphoricarpos* comes from the two Greek words *symphorein*, bunched together, and *karpos*, fruit, in reference to the clusters of white berries. *Mollis*, *albus*, and *laevigatus* are Latin words meaning soft, white, and smooth.

Poison Oak
Toxicodendron diversilobum (T. & G.) Edward Greene (1843-1915)
Anacardiaceae *(tock-sih-koe-DEN-drawn dye-ver-sih-LOE-bum)*

Any person living in California who is remotely interested in the outdoors should learn to recognize poison oak, one of our most hazardous native plants. It is a deciduous, perennial shrub, or sometimes a large, woody, climbing vine, with leaves that are divided into three leaflets that range in color from bright green to dark red. Being one of the most widely distributed native plants in California, it is common in many different habitats and often dominates the understory in coastal oak woodlands. When the leaves and stems come in contact with the skin, a blistering, itching rash can form. All parts of the plant contain urushiol, the toxic substance in the sap that causes an allergic skin reaction in most people. Firefighters are often affected by urushiol in the smoke from burning poison oak, and many who battle summer blazes in California are so severely affected that they are unable to work. Although it is called an oak, it is not a true oak, and is a member of the same plant family as poison ivy (*T. radicans*), cashews (*Anacardium occidentale*), mangos (*Mangifera indica*), and pistachios (*Pistacia vera*). *Toxicodendron* means poison tree, *diversilobum* means variably lobed, describing the variability of the three separated leaf lobes or leaflets.

California Bay-Laurel
Umbellularia californica Thomas Nuttall (1786-1859)
Lauraceae *(um-bell-yew-LARE-ee-ah kal-lih-FOR-ni-kuh)*

The California bay-laurel is an evergreen tree ranging from Baja California to Southwest Oregon. It is commonly associated with sycamores and oaks in shaded riparian and oak woodland areas. The California bay-laurel bears small yellowish-green flowers in umbel clusters from December to April, which are followed by olive-like fruits that turn purple when ripe. This tree is the only California native in the mostly tropical Laurel family, which gives us avocados, cinnamon, and camphor. The fine-grained, yellow-brown wood, sometimes called Oregon myrtle, is prized for its quality and is used in furniture, bowl turning, and carving. The dark green, oily, pungent leaves can be used for flavoring and insect repellent and a tea from these leaves has been used to treat rheumatism, stomach problems, and headaches, although the smell can also be a source of headaches. The name *Umbellularia* means little umbrella in Latin, in reference to the shape of the flat-top floral clusters.

Lichens

(LYE-kens)

Although not technically plants, lichens are present anywhere plants grow, just look closely at any nearby rock or tree trunk. Thousands of species of lichens are found in every habitat around the world, including a wide variety in the San Luis Obispo area. These amazing organisms are actually a living association between fungi and photosynthetic algae or bacteria. The fungus provides the body, while the smaller, photosynthetic symbiont supplies food. Depending on their shape, lichens are categorized as crustose (crusty), foliose (leafy), or fruticose (bushy).

Other oak woodland and riparian plants. Clockwise from top left. California figwort (*Scrophularia californica*), leather root (*Hoita macrostachya*), columbine (*Aquilegia eximia*), and leopard lily (*Lilium pardalinum* ssp. *pardalinum*).

The Sand Dunes

Life on the beach can be harsh on plants. Numerous hazards need to be overcome on a daily basis: the possibility of being submerged in salt water; low-nutrient, salty, sandy soil that is constantly moving; a temperature that is at times very hot and at times very foggy and cool; constant wind and salt spray; and regular drought. It is amazing that there are any plants living on the dunes at all. However, in spite of the tough conditions, a mosaic of different species blankets this unique habitat.

These plants have special characteristics that allow them to live in this challenging environment: most stay low to the ground, forming dense mats; many have succulent, water-storing leaves and deep anchoring tap roots; some excrete salt; and others have short life spans and buoyant seeds. Many of these plants are able to affect the dune environment and begin to slow the windblown sand, eventually forming a mound around them called a hummock. If numerous plants are able to colonize an area, the sand can become completely stabilized and covered with vegetation. The stabilization of sandy soils is a precarious business however, which can be reversed by disturbances both large and small. Storms, fires, off-road vehicular activity, and even foot traffic can cause blowouts in which sand begins to move again, engulfing established areas in its path. For this reason, efforts have been made to restrict access to sensitive dune habitats. The following pages contain descriptions of species that dominate the dunes.

Sand Verbena

***Abronia* spp.** Antoine Laurent de Jussieu (1748-1836)
Nyctaginaceae (*uh-BROE-nee-ah*)

There are three different sand verbenas that are common on the Central Coast dunes. These include the two pictured above, yellow sand verbena (*A. latifolia*) and purple sand verbena (*A. umbellata*), and the species pictured on the left, beach sand verbena (*A. maritima*), with wine-colored flowers. Like many plants on the harsh, windswept dunes, sand verbenas are low and trailing. The umbrella-shaped floral heads bloom from May through November. All three species can hybridize with each other, and if you look closely, you can find plants with intermediate characteristics. Sand sticks to the thick, glandular leaves and flowers, coating and protecting them from abrasion. The name *Abronia* comes from the Greek word *abros*, meaning delicate or graceful.

Beach-bur
Ambrosia chamissonis (Lessing) Edward Greene (1843-1915)
Asteraceae (*am-BROE-zi-ah kam-mih-SOE-niss*)

Beach-bur is an interesting member of the sunflower family that colonizes unstable or moving sand dunes. It is one of a small group of pioneer plants found in an area of actively moving dunes immediately adjacent to the beach, sometimes called the foredunes. This sprawling perennial is native to beaches and dunes from Baja California all the way north to British Columbia. Its gray, fragrant leaves, which vary in shape from slightly toothed to completely lobed, are covered with fine hairs. In the mid to late summer, the tips of certain stems develop spike-like clusters of tiny inflorescence heads. The upper portion of these inflorescences has only male flowers clustered together into small heads while single female flowers, which develop into spiny, bur-like fruit, are gathered below (see opposite page photo). While naming this species, the German physician Christian Lessing was honoring Adelbert von Chamisso, a Russian botanist who visited the west coast in the early 1800s and was the first to collect and name a number of California's plants. Plant names are not static, and Edward Greene later moved this species from Lessing's original genus, *Franseria*, to *Ambrosia*. The genus *Ambrosia*, which also contains the ugly, rank-smelling, allergenic ragweeds, was given its name, the Greek name for the food of the Gods, by Carolus Linnaeus. This name has led some botanists to wonder if he was making a joke.

Sea Rocket
Cakile maritima Giovanni Scopoli (1723-1788)
Brassicaceae (kah-KEE-lee mah-RIH-tih-mah)

This non-native member of the mustard family can be found on beaches, dunes, and sandy areas just above the high-tide line, from Baja California north to British Columbia. It is a small herb with succulent, lobed, hairless leaves, and flowers with four green sepals, four cross-shaped, pink petals, and six yellow stamens. Pollination of these flowers leads to the development of inch-long rocket-shaped fruits. These fruits are divided by an upside down V-shaped line, which breaks apart when dried. The single seed above this line stays in the corky fruit and is dispersed by floating in the waves; the seed below the line remains attached to the mother plant, which is buried by shifting sands in an environment already proven suitable. These two methods for dispersal increase the chances of survival of the next generation and have helped sea rocket, a native of Europe and North Africa, spread up and down our coastline. *Cakile* is the Arabic name for this plant, and *maritima* means of the sea.

Beach Primrose
Camissonia cheiranthifolia (Sprengel) Rudolf Raimann (1863-1896)
Onagraceae (*kam-ih-SOE-nee-ah kurr-an-thih-FOE-lee-uh*)

The beach primrose is a member of the large sun cup genus, *Camissonia*, which has over forty species living in California. It is an important dune stabilizer and can be found on upper beaches and coastal dunes from southern Oregon throughout most of coastal California to northern Baja California, blooming from March until early summer. This species grows as a low, sprawling, perennial herb or subshrub with thin trailing stems radiating from a rosette of thick, densely hairy, grayish-green leaves. The closely packed leaf hairs help this plant survive desiccation and sand abrasion caused by the constant ocean breeze. The bumblebee-pollinated flowers, which often face downwind and are open only during the day, have four bright yellow petals that turn red with age. This genus honors the person who named the California poppy, Adelbert von Chamisso. He was the naturalist on the Russian ship the *Rurick*, which visited California in 1816. The species name indicates that this plant's leaves are like those of mustard family wallflowers in the genus *Cheiranthus*, which has now been renamed.

Ice Plant
***Carpobrotus* spp.** (Linnaeus) Nicholas Brown (1849-1934)
Aizoaceae *(kar-poe-BROE-tuss)*

Ice plant is a non-native weed that forms wide, dense mats, often at the expense of native dune vegetation, in coastal areas throughout California and Baja California. The two ice plant species, *C. edulis* (Hottentot fig) and *C. chilensis* (sea fig), which are both native to South Africa, often hybridize. *C. edulis*, which has been extensively planted for erosion control along Southern California highways, has a slightly toothed outer edge on its three-sided succulent leaves. Its yellow flowers, which sometimes turn pink with age, are larger than those of *C. chilensis*. *C. chilensis*, whose magenta flower is pictured here, was once thought to possibly be native to California and is less invasive than *C. edulis*. *Carpobrotus* comes from the Greek words *karpos*, meaning fruit and *brotos*, edible, referring to the many-seeded berry, which was eaten by aboriginal South Africans.

Slender-leaved Ice Plant
Conicosia pugioniformis (Linnaeus) Nicholas Brown (1849-1934)
Aizoaceae (*koe-nih-KOE-zsa pew-JEE-uh-nih-for-miss*)

Slender-leaved ice plant is a native of South Africa that has become a well-established, invasive weed on California dunes from San Francisco to northern Santa Barbara County. It is particularly abundant on the Nipomo-Guadalupe dunes and Vandenberg Air Force Base. This short-lived perennial is easily distinguished from its other ice plant (*Carpobrotus* spp.) relatives by its long slender leaves that grow in ground level clumps and rosettes, as opposed to the thick leaves and mat forming shoots of *Carpobrotus*. This plant's grayish-green, hairless, succulent leaves are triangular in cross section. The ephemeral, unpleasant-smelling, solitary flowers emerge from spring to fall and have five succulent sepals whorled around many shiny, papery, bright yellow petals. Inside the petals are numerous stamens and an ovary that develops into a cone-shaped capsule, which splits open when dried. The constant sea breeze blows these dried fruits away from the mother plant, which disperse their seeds as they roll. The name of this genus is derived from the fruit shape; *Conicosia* is Greek for cone-shaped. *Pugioniformis*, which means dagger-like in Latin, refers to the narrow, pointed leaves.

California Croton
Croton californicus Jean Mueller (1828-1896)
Euphorbiaceae (*KROE-tawn kal-lih-FOR-ni-kuss*)

California croton is a low-growing perennial or subshrub, rarely taller than a couple feet, with gray leaves. It is common in the sandy soils of stabilized dunes, comprising part of the understory of the coastal dune scrub plant community. It ranges natively along the coast from San Francisco south to Baja California. This species is dioecious, meaning flowers of some plants are male, with stamens, while flowers of other plants are female, with pistils. Once female flowers are pollinated, the pistil gives rise to a three-sided capsule fruit that explodes when fully dry. *Croton* is derived from *kroton*, the Greek word for a tick. Seeds of many members of this family are adorned with a small head-like outgrowth called a caruncle, giving the seeds the appearance of an engorged tick (see castor bean (*Ricinus communis*) discussion on page 142).

Veldt Grass
Ehrharta calycina James Smith (1759-1828)
Poaceae (*air-HAR-tuh kal-lih-SEE-nuh*)

In many areas of sand dune habitat, the human impact on native plants has been severe. Most of these stubbornly resilient plants, which are adapted to survive constant wind, salt spray, and moving sand, usually cannot survive being trampled, run over, or out-competed by exotic invasive species such as veldt grass. A South African native, veldt grass was brought to California around 1930, after first being introduced to Australia, because it grew well in sandy soils and could be used for erosion control and livestock forage. It has now become a wildly aggressive weed on the dunes, displacing much of the native vegetation where it grows. Veldt grass is a perennial, bunch-forming grass that forms almost pure stands. It can be easily recognized while flowering in spring when it forms two- to three-foot tall, rust-colored inflorescences. This genus was named in honor of the eighteenth-century Swiss botanist J. Friedrich Ehrhart, a student of Carolus Linnaeus. *Calycina* means calyx-like, possibly referring to the two, small, rust-colored, specialized leaves that surround the tiny flowers.

Mock Heather
Ericameria ericoides (Lessing) Willis Jepson (1867-1946)
Asteraceae (*air-rih-kuh-MAIR-ree-uh air-rih-KOE-ee-dees*)

Mock heather is an evergreen, compact shrub with a native range from Marin County south to Los Angeles. It can occasionally be found on the outer dunes, but more frequently lives on stabilized soil as a dominant member of the coastal dune scrub plant community, just inland from areas of actively moving sand. It makes fan-shaped clusters of small, resinous, needle-like leaves, each with a shallow groove on the back. Mock heather becomes very showy in the late summer and fall when it produces bright yellow ray and disk flowers in a profusion of compact heads. Willis Linn Jepson, the founder of the California Botanical Society and an early twentieth-century botany professor at U.C. Berkeley, named this species. Jepson, who was the first to publish a comprehensive manual of California plants, was the most prominent plant taxonomist in the state for many years. His legacy still lives on with the 1993 publication of *The Jepson Manual: Higher Plants of California*; an ambitious collaborative project with 186 authors that now serves as the standard manual for identifying California plants. *Ericameria* means having parts like *Erica*, which is the genus of the heathers, important horticultural plants that also have small needle-like leaves. *Ericoides* means resembling *Erica*.

Coastal Buckwheat
Eriogonum parvifolium James Smith (1759-1828)
Polygonaceae (*air-ee-AW-goh-num par-vih-FOE-lee-um*)

The wild buckwheat genus, *Eriogonum*, is one of the most diverse in California. There are about 250 species in North America, half of which are found in California. Coastal buckwheat, one of San Luis Obispo County's twenty buckwheat species, only extends inland a few miles and can be found on stabilized dunes, sandy flats, and rocky coastal slopes from Monterey County south to San Diego. It is a low-growing, woody shrub whose leaves have white, woolly undersides and often turn red with age. The white or slightly pink flowers are borne in puffy heads throughout the summer. *Eriogonum* comes from two Greek words, *erion*, meaning woolly and, *gony*, a knee, referring to the woolly nodes of some of the first species ever named in this genus. *Parvifolium* means small leaves.

Other plants on the dunes. Above clockwise from top left. Beach morning glory (*Calystegia soldenella*), dunedelion (*Malacothrix incana*), horkelia (*Horkelia cuneata*), Blochman's leafy daisy (*Erigeron blochmaniae*), sand strawberry (*Fragaria chiloensis*), and crisp dune mint (*Monardella crispa*). Opposite clockwise from top left. Dune wallflower (*Erysimum insulare suffrutescens*), Indian pink (*Silene laciniata*), cob web thistle (*Cirsium occidentale*), and purple sand food (*Pholisma arenaria*).

Grasslands & Wildflowers

Grasslands are areas dominated by grasses and non-grassy, herbaceous plants called forbs. Grasslands occur on rich soils in flat areas and rolling hills throughout California. California grasslands have changed so drastically in the last 200 years that now any typical grassland hillside is most likely a lawn of weeds. Native, perennial, bunch-forming grasses dominated California grasslands until the time of the earliest Spanish incursions when they began to be displaced by weedy annual grasses from Europe. In fact, the nearly complete replacement of native California grasslands by those dominated by weeds is one of the most dramatic examples of ecological invasion in recent history. Periods of drought, overgrazing, changes in fire frequency, as well as displacement by farming and development have all favored growth of weedy, exotic grasses and forbs, and now only scattered, relic areas of native grasslands remain.

Many grassland forbs, or wildflowers, make spectacular spring blooming shows. In some inland areas of the Central Coast, such as Shell Creek and the Carrizo Plain, magnificent wildflower blooms can create unbroken carpets of color. The following pages describe some of the most common native and non-native grasses and the most conspicuous and abundant wildflowers in the San Luis Obispo area.

Fiddleneck
Amsinckia spp. Johann Lehmann (1792-1860)
Boraginaceae *(am-SINK-ee-ah)*

Most of the approximately ten species of fiddlenecks are native to the Central Coast. These annual wildflowers are abundant in spring on grassy hillsides, in recently burned areas, and along roads. They are easily recognized by their long, golden, uncoiling, floral clusters, with a shape reminiscent of the neck of fiddle. All species have small tube-shaped flowers that are orange, yellow, or some combination of the two colors. The leaves and stems are covered with bristly hairs and can cause irritation when touched by people with sensitive skin. Some species are also toxic when eaten by livestock. The two species pictured here are the common fiddleneck (*A. menziesii*, left) and the seaside fiddleneck (*A. spectabilis*, right). Johann Christian Lehmann, an early nineteenth-century director of the Hamburg Botanical Garden and professor at Hamburg University, named this genus in honor of Dr. Wilhelm Amsinck, a rich patron of the gardens.

Wild Oats
***Avena* spp.** Carolus Linnaeus (1707-1778)
Poaceae *(ah-VEE-nah)*

Wild oats are ubiquitous, non-native, annual grasses in California, and some of the worst weeds in the world. Like many of the other non-native Mediterranean grasses, wild oats have aggressively and successfully out-competed California's perennial, bunch-forming grasses in many local and statewide grasslands. The disturbing aspect of the hegemony of weedy grasses such as wild oats is that they appear to inhibit any attempts by native grasses to recover a position in the grassland community, even in the absence of grazing. Unfortunately, the truth may be that these invaders are here to stay. There are three species of *Avena* that occur frequently on the California Central Coast, common wild oats (*A. fatua*), slender wild oats (*A. barbata*), and the non-weedy cultivated oats (*A. sativa*). The two weedy species can be distinguished by the presence (*A. barbata*) or absence (*A. fatua*) of two hair-like bristles at the tip of the lemma (the modified leaf enclosing the flower). *Avena* is the classical Greek name for cultivated oats, which were domesticated by early Mediterranean farmers around 4,000 years ago.

Ripgut Brome
Bromus diandrus Albrecht Roth (1757-1834)
Poaceae *(BROE-muss dye-AN-druss)*

There are approximately thirty brome grasses (*Bromus* spp.) in California, half of which are non-native naturalized weeds. Some of these, like ripgut brome (two pictures on the left), with stiffly barbed awns (long bristles extending from the floral heads) can be very irritating and injurious for livestock. Hold the tip of an awn with one hand and pull downward on it with your free thumb and forefinger to get an appreciation for why this grass is named ripgut. It can be found in disturbed areas throughout lower elevation regions of western North America. *Diandrus* means having two stamens.

Red Brome
Bromus madritensis Carolus Linnaeus (1707-1778)
Poaceae *(BROE-muss mah-drih-TEN-siss)*

Red brome (three pictures on the right) is an annual weedy grass native to southern Europe that can pioneer the colonization of disturbed or burnt areas, sometimes diminishing in abundance if it is overrun by shrubs and other perennials. This grass has upright brush-like inflorescences that become wine colored in maturity and then fade to light tan as the summer progresses. *Bromus* is derived from, *bromos*, the ancient Greek word for oats, and *madritensis* possibly refers to the first botanical collection of this species in Madrid, Spain.

Mariposa Lilies
***Calochortus* spp.** Frederick Pursh (1774-1820)
Liliaceae *(kal-loe-KORE-tuss)*

There are sixty species of *Calochortus*, all occurring in western North America, from British Columbia south to Guatemala. It is a real treat to come upon a blooming mariposa lily on a summer walk, as they are among the most beautiful wildflowers in San Luis Obispo. They form underground bulbs, which were dug up and eaten raw or roasted by aboriginal Californians. During much of the year, before blooming, they can be recognized only by one large strap-like leaf emanating from the soil. All mariposa flowers have three petals, which are often vividly colored with blotches, covered with dense hairs, and have a pollination nectary near the base of each petal. The name *Calochortus* is Greek for beautiful grass. The common name mariposa means butterfly in Spanish, probably referring to the appearance of the large spotted flowers floating above the grass in a gentle wind. The species pictured here are (clockwise from top left) the white fairy lantern (*C. albus*), the yellow mariposa (*C. clavatus*), the butterfly mariposa (*C. venustus*), and the San Luis star-tulip (*C. obispoensis*), which is listed by the California Native Plant Society as being rare and endangered, existing nowhere else but around San Luis Obispo.

Indian Paintbrush
Castilleja affinis William Hooker (1785-1865) & George Arnott (1799-1868)
Orobanchaceae *(cass-still-LAY-uh AFF-ih-niss)*

Indian paintbrush is a branched perennial that usually grows less than two feet tall. It is native to woodlands, coastal scrub, and chaparral in both coastal and interior lower elevation areas of most of western California. It is closely related to purple owl's clover (*Castilleja exserta*), but belongs to a different group of species with mostly reddish flowers within the same genus. In spring and early summer Indian paintbrush blooms with spike-like clusters of numerous, inconspicuous, tubular flowers surrounded by large, brightly colored, red and yellow bracts. This species is divided into a number of subspecies that differ in native range and morphology. The one shown here, and the most common, is *C. affinis*, ssp. *affinis*. Indian paintbrush has green leaves, indicating that it produces its own food by photosynthesis. However, a close investigation of its roots reveals this plant's partially parasitic nature. It extracts nutrients from the roots of nearby shrubs, a characteristic making this and other related species difficult to transplant or grow from seed outside their natural habitat. The genus was named in honor of the Spanish botanist Domingo Castillejo, and *affinis* means bordering, related to, or similar to.

Purple Owl's Clover
Castilleja exserta
(Heller) Tsan-Lang Chuang (1933-1994) & Lawrence Heckard (1923-1991)
Orobanchaceae *(cass-still-LAY-uh ek-SUR-tuh)*

Purple owl's clover was originally a member of the genus *Orthocarpus* until the early 1990s when it was moved to the large genus *Castilleja*, named in honor of Domingo Castillejo, an eighteenth-century Spanish professor of botany. Owl's clover is a native annual species (grows from seed each year) that is partially parasitic on the roots of other plants. It occurs in open grasslands, sometimes covering areas with beautiful magenta patches during spring blooms, throughout much of California, Arizona, and northern New Mexico. It has a spike-like inflorescence containing many small tubular flowers each sitting just above a tiny, lobed, colorful bract leaf. From a distance owl's clover has the superficial appearance of clover, but a careful look easily separates the two. Upon a close examination of the small flowers one can see the velvety, narrow, hooded, upper corolla. Just below this hooked structure is the puffy lower corolla lip, which can sometimes look like the face of an owl (see inset upper left photo). *Exserta* means protruding in Latin, possibly referring to the projecting floral organs.

Soap Plant
Chlorogalum pomeridianum
Augustin Pyramus de Candolle (1778-1841)
Liliaceae *(KLOR-roe-gall-um pawm-mer-rih-DEE-an-um)*

Soap plant is common in grassland areas throughout California. Its long, wavy, blue-green leaves emanate from a bottle-shaped underground bulb that is covered with brown fibers. Aboriginal Californians put soap plant to use in many ways. The bulb fibers were used for mattress stuffing and making fine brushes, the bulb was a source of soap in addition to being roasted and eaten. Mashed raw bulbs were scattered in dammed streams to stupefy fish, and the bulb's sticky mucilage was used as glue for arrow feathers. This genus takes its name from the Greek words *chloros* and *gala*, meaning green milk, referring to the plant's green sap. *Pomeridianum* is Latin for postmeridian, referring to the fact that this plant's small, star-shaped, white flowers only open late in the afternoon before being pollinated by insects in the evening.

Clarkias

***Clarkia* spp.** Frederick Pursh (1774-1820)
Onagraceae *(clar-KEE-uh)*

The large *Clarkia* genus, with over forty species in California, was named for William Clark, one of the two leaders on the Lewis and Clark expedition across the northwestern United States, which took place 200 years ago. Thomas Jefferson sent Lewis and Clark to explore uncharted areas of the Louisiana Purchase and to try to find a water route from the Mississippi to the Pacific Ocean. Along the way, they discovered many plant species that had never been described botanically. Wildflowers in the *Clarkia* genus, which are some of the last to bloom in spring and often go by the name farewell-to-spring, have brilliantly colored flowers ranging from white or pale pink to lavender or burgundy. They grow in open grasslands, chaparral, coastal scrub, and wooded areas throughout western California. *Clarkia* seeds were collected by aboriginal Californians and eaten alone, with acorn meal, or in pinole, a finely ground flour made from the seeds of grasses and wildflowers, which was eaten dry or combined with water to make a mush. The species pictured here are the elegant clarkia (*C. unguiculata*, left) the punch-bowl godetia (*C. bottae*, upper right), and the four spot (*C. purpurea* ssp. *viminea*, lower right).

Larkspur
***Delphinium* spp.** Carolus Linnaeus (1707-1778)
Ranunculaceae *(dell-FIN-ee-um)*

There are approximately ten larkspurs that occur in the San Luis Obispo area. The one pictured here (left photo) is one of the most common, Parry's larkspur (*Delphinium parryi*). A distinctive feature of larkspur flowers is the upper, showy, petal-like sepal that forms a spur extending horizontally behind the flower. Some species are extremely toxic to cattle and can cause death from paralysis if enough is eaten. *Delphinium* is derived from the Latin word *delphis*, a dolphin, referring to the shape of the flower.

Blue Dicks
Dichelostemma capitatum (Bentham) Alphonso Wood (1810-1881)
Liliaceae *(dih-keh-low-STEM-mah kah-pih-TAY-tum)*

Blue dicks are found throughout the Western U.S. and Northern Mexico growing in a variety of habitats. The clustered flowers, which range from dark purple to light pink, radiate from the top of a leafless stem that arises from an edible, underground corm (a bulb-like structure surround by papery leaves). Usually two or three long, grass-like leaves can be found at the base of the floral stem. *Dichelostemma* is Greek for a twice-parted crown, referring to the wing-like appendages covering half the stamens. *Capitatum* refers to the flowers growing in a head.

Chocolate Lily

Fritillaria biflora John Lindley (1799-1865)

Liliaceae *(frih-tih-LARE-ee-uh BYE-floor-uh)*

The dark brown, bell-shaped flowers of the chocolate lily nod downward on the stem. These beautiful members of the lily family bloom in open grassy areas in early spring. *Fritillaria* comes from the Latin word *fritillus*, a dice box, possibly referring to the spotted markings on the flowers or the general shape of the flower. The species name *biflora* alludes to fact that often only two flowers are borne on each plant.

Shooting Star

Dodecatheon clevelandii Edward Greene (1843-1915)

Primulaceae *(doe-deh-KAH-thee-on kleev-LAN-dee-eye)*

The leaves of this unforgettable wildflower are whorled at the base of the flowering stem. At the tip of the stem are a number of downward facing rocket-shaped flowers with backward-bent, pink petals exposing two rings of yellow and white at their base. These spring blooming flowers start by facing upward, turn downward as they develop, then turn upward again after they are pollinated. *Dodecatheon* means twelve gods, first used by the Roman naturalist Pliny to describe a flower thought to be protected by deities. *Clevelandii* commemorates Daniel Cleveland, a plant-collecting attorney who started the San Diego Natural History Society.

California Poppy
Eschscholzia californica Adelbert von Chamisso (1781-1838)
Papaveraceae *(esh-SHOLL-zee-ah kal-lih-FOR-ni-kuh)*

One of the most widespread wild flowers in the state, and the official state flower, the California poppy can be found growing on open grassy areas, sand dunes, and disturbed sites throughout California. This low-growing annual has four beautiful, satiny, golden orange (inland) or yellow (coastal) petals that form an upright cup. As early as January and as late as September, these petals emerge as they push off the two cone-shaped, protective sepals. As the cylindrical capsule fruit dries it can forcibly and audibly pop while dispersing the seeds within. Only after cooking were the toxic stems and leaves of California poppy eaten by aboriginal Californians. Adelbert von Chamisso was the naturalist on a Russian exploration trip to the Pacific coast in 1816 when he named the California poppy in honor of J. F. Eschscholtz, the ship's surgeon and entomologist. In publishing the genus, Chamisso omitted the t from Eschscholtz's name. The first collected specimen is in Leningrad, Russia.

Foxtail Barley
Hordeum murinum Carolus Linnaeus (1707-1778)
Poaceae *(HOR-dee-um MIR-eye-num)*

Foxtail barley, sometimes called hare barley, is an annual, weedy grass imported by the early Spanish settlers from the Mediterranean region. It is now distributed throughout most of the United States where it can be found growing on roadsides, in and near pastureland, fencerows, and ditches. In California this species grows vigorously during the winter, flowers in spring, and dries up, living only as dormant seeds during the summer. The smallest unit of a grass inflorescence is called a spikelet. The numerous spikelets in a foxtail barley inflorescence are all armed with long, stiff, bristle-like appendages called awns, giving the entire inflorescence a bushy appearance. Once the grass becomes dry, the inflorescences easily falls apart and the barbed awns on the spikelets can become an annoyance for animals who get them stuck in their fur, nostrils, eyes, ears, and throat. *Hordeum* is the ancient Latin name for cultivated barley (*H. vulgare*) and *murinum* means mouse-like, possibly in reference to the bushy inflorescences.

Goldfields
Lasthenia californica John Lindley (1799-1865)
Asteraceae *(lass-THEN-nee-ah kal-lih-FOR-ni-kuh)*

Goldfields is an abundant annual wildflower in the sunflower family that rarely reaches heights over six inches. This species can be found growing in moist, flat, grassy areas from the immediate coast throughout most of western California into southwestern Oregon and eastward to Arizona. In March and April, goldfields lives up to its common name by blanketing large areas with rich, golden blooms. Individual flower heads are no more than an inch across and have the typical sunflower-like ray and disk flowers compacted into heads (see the sunflower discussion in the tidy tips description on the following page). Some species of *Lasthenia* are common in areas called vernal pools. These are depressions in flat grasslands that collect winter rains, creating a shallow pool for a short period before completely drying in the late spring. This special, ephemeral habitat, which supports numerous endemic plant and animal species, is becoming exceedingly rare in California due to development and agriculture. *Lasthenia* is named in honor of a Greek girl by the same name who disguised herself as a boy in order to attend Plato's lectures.

Tidy Tips
Layia platyglossa (Fischer & Meyer) Asa Gray (1810-1888)
Asteraceae *(LAY-ee-ah plah-tee-GLOSS-uh)*

Tidy tips are annual wildflowers in the sunflower family, the largest plant family in California and the world's second largest plant family (the largest is the orchid family [Orchidaceae] with nearly 30,000 species). Sunflowers, daisies, and their many relatives such as tidy tips are not single flowers but specialized clusters of many small flowers appearing in radially symmetrical, complex heads. At a distance, and probably to their pollinators, these inflorescences (floral clusters) appear to be one large flower. Minute disk flowers reside in the center of the head, while the petal-like ray flowers are arranged around the periphery. Tidy tips have large yellow ray flowers with creamy white tips, described by the species name, which means broad tongue. They bloom in March on coastal dunes and in inland valleys, sometimes forming large carpets of color. The genus was named for George T. Lay, an early nineteenth-century English plant collector who visited California in 1827.

Italian Ryegrass
Lolium multiflorum Jean Lamarck (1744-1829)
Poaceae *(LOE-lee-um mull-tee-FLOOR-um)*

There are eight species of *Lolium*, all native to Northern Africa, Western Asia, and the Mediterranean region. One of these, Italian ryegrass, was brought to California with the first European settlers in livestock hay and has since become an important, if not ubiquitous and noxious, naturalized component of the state's grasslands. It is an annual, sometimes biennial, weedy grass that has spread throughout the U.S., growing on grassy slopes, abandoned fields, meadows, and disturbed sites. *Lolium* can be recognized by its spikelets, the flower containing unit in the inflorescence, which are arranged in two opposite ascending ranks along the stem. This species can be distinguished from others in the genus by the fact that it is not perennial and has short awns (bristle-like extensions) on it spikelets. Pollen released from these flowers every spring causes great distress for sufferers of hay fever, as ryegrass pollen is hyperallergenic. The Roman poet Virgil (70-19 B.C.) is credited with first using the name *Lolium* to describe a weedy grass, and *multiflorum* means many flowered.

Lupines
***Lupinus* spp.** Carolus Linnaeus (1707-1778)
Fabaceae *(LEW-pye-nuss)*

There are over twenty species of lupines growing in the San Luis Obispo area, and over eighty in California. Among the most recognizable of wildflowers, they bloom in spring in all California plant communities except salt marshes. Lupines range from annual and perennial herbs to large shrubs. The pea-shaped flowers, which come in a variety of colors, have five petals arranged in a specific orientation. One petal, called the banner, stands upright, two side petals make up the wings, and the two remaining petals are fused on the lower side forming the keel. All lupines have palmately compound leaves, with leaflets emanating from the tip of the leaf stalk (petiole), like fingers from the palm of a hand. The generic name, from the Latin word *lupus*, meaning wolf, was mistakenly applied to these plants because they were thought to rob the soil of nutrients, like a wolf robbing a farmer's chickens. However, due to a symbiotic relationship with nitrogen-assimilating bacteria in their roots, they do quite the opposite. The species pictured here are (clockwise from top left) silver lupine (*L. albifrons*), stinging lupine (*L. hirsutissimus*), arroyo lupine (*L. succulentus*), silver dune lupine (*L. chamissonis*), the arroyo lupine inflorescence (inset), and sky lupine (*L. nanus*, opposite page).

Purple Needle Grass
Nassella pulchra (Hitchcock) Mary Barkworth (1941-)
Poaceae *(nah-SELL-ah PULL-krah)*

Three hundred years ago, before the first Europeans brought their weedy grasses and domesticated livestock to California, a visitor to the state would have seen hillsides and grasslands dominated by perennial bunch-forming grasses such as purple needle grass. After years of drought, overgrazing, weed invasion, and fire suppression, native grasslands have been reduced to only remnant stands scattered throughout the state. Purple needle grass, which was one of the dominant grasses in the foothills surrounding the Central Valley and throughout the Central Coast, still has a similar range, but with much reduced populations. It has two- to three-foot flowering stems with deep purplish-brown flowers arranged in an open, spreading, often drooping inflorescence. This species can be distinguished from the two other related species in the San Luis Obispo area (*N. cernua* and *N. lepida*) by its one- to four-inch long, needle-like awns (bristle-like extensions on the spikelets), which are bent twice then straight at the tip. Purple needle grass is now the official California state grass, signed into law by Governor Schwarzenegger. This genus has a name derived from the Latin word *nassa*, a basket with a narrow neck, possibly referring to the spikelet narrowing to a pointed awn. *Pulchra* means beautiful.

Baby Blue-Eyes
Nemophila menziesii
William Hooker (1785-1865) & George Arnott (1799-1868)
Boraginaceae (nee-MOFF-fill-luh men-ZEE-see-eye)

Baby blue-eyes is a common and well-known annual wildflower that blooms early in the spring. Its delicate stems sprawl low to the ground with two slightly hairy, lobed leaves attached at each node. The large bowl-shaped flowers have bright blue petals with a white center, sometimes marked with dark lines or spots. This species grows in meadows, shrublands, roadsides, and woodlands throughout California. Baby blue-eyes is a popular garden flower in Europe since being introduced from California in the 1830s by David Douglas, a Scottish plant collector for the Royal Horticulture Society. *Nemophila* comes from the Greek words *nemos*, a wooded meadow, and *phileo*, to love, in reference to an affinity for woodlands. The species name commemorates Archibald Menzies, a Scottish naturalist and physician aboard Captain George Vancouver's ship during an exploration of the Pacific that included a visit to California in the late 1700s. Many western North American and Hawaiian plants are named after Menzies, who is credited with discovering the California state tree, the coastal redwood (*Sequoia sempervirens*).

Phacelias

***Phacelia* spp.** Antoine Laurent de Jussieu (1748-1836)
Boraginaceae *(fah-SEE-lee-ah)*

These wildflowers are all members of the large (nearly 200 species), entirely New World genus *Phacelia*. There are over 100 species in California and more than twenty on the Central Coast, often making the casual recognition of individual species rather difficult. *Phacelia* is from the Greek word for a bundle, *phakelos*, a name used for this genus to describe the densely crowded flowers, which are usually clustered on one side of a coiled inflorescence called a cyme. In this type of inflorescence the middle or upper flowers open first, a feature easily seen on most phacelias. Individual flowers, which range in color from white to dark purple, are bell shaped with petals flaring outward. The leaves are often pinnately lobed or compound (with leaves being divided into smaller leaflets). The bristly, sticky hairs of some species can cause staining or rashes on the skin when touched. The species picture here are (clockwise from top left) *P. brachyloba*, *P. distans* (common phacelia), *P. imbricata* (imbricate phacelia), and *P. grisea* (windowed phacelia).

California Buttercup
Ranunculus californicus George Bentham (1800-1884)
Ranunculaceae (rah-NUN-kew-luss kal-lih-FOR-ni-kuss)

Ranunculus is a large genus of over 600 species native to temperate areas worldwide. Many are cultured ornamentally and all contain the poisonous glycoside ranunculin. California buttercup, the most common and widespread *Ranunculus* in the state, is an early-flowering, annual wildflower. It has shiny one-inch-wide yellow flowers at the ends of branched inflorescences, rising from deeply-divided, basal leaves. After pollination, the many pistils at the center of the flower develop into little hooked achenes (one-seeded fruits), which were dried and eaten by aboriginal Californians. *Ranunculus* means little frog, a derivation of the Latin word for frog, *rana*, a name given to this genus possibly because many species share their moist habitats with frogs. George Bentham, British taxonomist and long-time president of the Linnaean Society of London, named this species as well as many other common California plants.

Blue-Eyed Grass
Sisyrinchium bellum Sereno Watson (1826-1892)
Iridaceae *(siss-ih-RIN-kee-um BELL-um)*

Blue-eyed grass is a common perennial herb found on moist grassy slopes throughout California. Like its relatives the garden irises, blue-eyed grass has underground stems and long, flat, basal leaves. Although called a grass due to its superficial grass-like appearance, it is a member of the Iris family. The six-parted, intensely bluish-purple flowers, which appear from April to June, have a bright yellow center, perhaps creating a large contrast that attracts pollinating insects. Spanish settlers used a tea made from the roots to cure fever. Theophrastus, a third-century B.C. student of Aristotle, first used the name *Sisyrinchium* to describe an Iris-like plant. This species was appropriately given the name *bellum*, meaning beautiful.

Clovers
***Trifolium* spp.** Carolus Linnaeus (1707-1778)
Fabaceae *(try-FOE-lee-um)*

The clovers of California are a very diverse group, with over thirty species, most of which are native. They are also relatively easy to recognize with their leaves divided into three toothed leaflets, a feature that gives the genus, *Trifolium*, its Latin name. These leaflets are usually arranged palmately, emerging from the tip of the leaf stalk (petiole) like three fingers from the palm of a hand. The small pea-like flowers occur in dense heads, umbels, or short spikes. Clovers have high nutritional content and aboriginal Californians ate the tender young leaves and seeds. They are also valued as forage, especially the non-native species. There are three native clovers pictured on the bottom (from left to right) *T. depauperatum* (dwarf clover), *T. wormskioldii* (Wormskjold's clover), and *T. bifidum* (Pinole clover), and three non-natives pictured on top (from left to right) *T. campestre* (hop clover), *T. arvense* (rabbitfoot clover), and *T. angustifolium* (narrow leaf clover).

Johnny-Jump-Up
Viola pedunculata John Torrey (1796-1873) & Asa Gray (1810-1888)
Violaceae *(vy-OH-lah pee-dunn-KEW-lah-tuh)*

The johnny-jump-up is a perennial wildflower with roundly triangular leaves on short above-ground stems arising from a rhizome (underground stem). It can be found on grassy slopes, in chaparral, and in oak woodlands of western California from Sonoma County south to Baja California. The orangish-yellow flowers appear in February with five petals, the lower three of which are decorated with dark brown to purple lines that probably serve as guides to direct pollinating insects. The insect eye is not efficient at discerning shapes but is exquisitely adapted to recognize contrasts in color. The flowers are borne on long stalks called peduncles, a distinguishing characteristic used when naming this species. *Viola* is an ancient name for small, sweet-smelling flowers. John Torrey and Asa Gray, whom collaborated on the original botanical description and naming of this and many other California species, were the two most important American plant biologist in their day. Torrey, at Columbia University in New York, and Gray, at Harvard University in Boston, received and studied thousands of plant specimens from explorers and plant collectors during the westward expansion of the United States in the middle of the nineteenth-century.

Other grassland plants and wildflowers. Above clockwise from top left. Golden stars (*Bloomeria crocea*), Chinese houses (*Collinsia heterophylla*), thistle sage (*Salvia carduacea*), and jewelflower (*Streptanthus glandulosus*). Opposite clockwise from top left. Slender-flowered gilia (*Gilia tenuiflora*), slender tarweed (*Deinandra fasciculata*), hayfield tarweed (*Hemizonia conjesta* ssp. *luzulifolia*), star lily (*Toxicoscordion fremontii*), Indian milkweed (*Asclepias eriocarpa*), checker bloom (*Sidalcea malviflora*), wind poppy (*Stylomecon heterophylla*), and melic grass (*Melica imperfecta*)

Grasslands and Wildflowers

Salt & Freshwater Marsh

Saltwater marshes, like those surrounding Morro Bay, occur in protected embayments scattered up and down the California coast. On a daily basis, the entire community becomes inundated with tidal saltwater, and the heavy clay soils are almost completely devoid of oxygen. These two environmental conditions require special characteristics of marsh plants. Relatively few plant species are capable of living in this harsh environment, and most are low-growing halophytes that mostly reproduce vegetatively. Saltwater marshes exist in areas of dense human population, and much of this important plant, marine, and bird habitat has been lost to dredging, infilling, and development.

Freshwater marshes exist in areas of standing or slow moving non-saline water. Many of the dominant freshwater marsh plants are grass-like monocots with special adaptations for living in water. Natural freshwater marshes are now restricted in their distribution compared to the days before agricultural infilling and draining began, when vast wetlands existed in the Central Valley. Today the aesthetic, environmental, economic, and spiritual value of healthy fresh and saltwater wetlands is beginning to be understood, and efforts are being made to preserve remaining patches. In the following pages, you will find descriptions of dominant salt and freshwater marsh species.

Saltbushes
***Atriplex* spp.** Carolus Linnaeus (1707-1778)
Amaranthaceae *(AH-trih-pleks)*

The forty California saltbushes are an extremely variable and apparently rapidly evolving group of species with many localized forms adapted to specific habitats. They are salt-tolerant, annual or perennial herbs and shrubs found in alkaline or saline soils. Saltbushes are common on the Central Coast on the borders of salt marshes, on dunes, beaches, and sea bluffs, and in inland deserts in salty, dry, alkali sinks. A saltbush can be recognized by its grayish or light green leaves, which often have a scaly surface covered with minute inflated leaf hairs, making them powdery and moist to the touch. These species are either monoecious or dioecious, meaning that their male and female flower parts (stamens and pistils) occur on the same individual plant, or on different individuals, but never within the same flower. Male flowers form inconspicuous masses of wind-pollinated stamens, while female flowers (without sepals, petals, or stamens) are pressed between two partially or wholly united scale-like bracts. Aboriginal Californians ate young shoots and leaves like spinach and ground seeds for meal. *Atriplex* is the ancient Latin name for saltbushes. The saltbushes pictured above, all of which are perennial natives, are Watson's saltbush (*A. watsonii*, background), big saltbush (*A. lentiformis* ssp. *lentiformis*, lower right), and California saltbush, (*A. californica*, top left). Spearscale (*A. triangularis*) is pictured on the opposing page.

Yerba Mansa
Anemopsis californica William Hooker (1785-1865)
Saururaceae *(an-nih-MOP-siss kal-lih-FOR-ni-kuh)*

Yerba mansa is a low-growing, creeping, perennial plant with basal heart-shaped leaves. It can be found in permanently wet, alkaline areas of Oregon, California, and eastward all the way to Texas. Around San Luis Obispo it resides in coastal marshes and inland alkali springs. What appear to be single flowers on this plant are actually highly modified spike inflorescences with many densely-clustered, petal-less flowers above modified leaves that look like white petals. The plant has a spicy odor, and a tea made from the leaves has been used for soothing wounds, purifying the blood, and treating colds, sore throats, dysentery, and asthma. Like many other highly aromatic plants, yerba mansa has been treated as a cure-all, although none of its uses have actually be scientifically substantiated. *Anemopsis* comes from the Greek words *anemone* and *opsis*, meaning anemone-like, referring to the similarity of the floral clusters to the large, showy flowers of the commonly cultivated anemone.

Sedges
***Carex* spp.** Carolus Linnaeus (1707-1778)
Cyperaceae *(CARE-eks)*

Sedges are perennial grass-like plants, associated with moist to wet habitats. *Carex* is an extremely large worldwide genus of around 2,000 species. With nearly 200 species in the state, this is California's largest plant genus. Although sedges can usually be recognized as such, the identification of different species is often difficult due to the minute differences distinguishing them. Typically sedge leaves are grass-like: long and slender with parallel veins, arising basally from an underground rhizome. The only above-ground stems, which are triangular in cross section, bear the inflorescences. These inflorescences range in shape from spike-like to head-like clusters of inconspicuous, petalless, wind-pollinated, unisexual flowers. Although sedge species are numerous and widespread, their human usage has been relatively minimal. Aboriginal Californians harvested rhizomes for basket weaving, and sedges are now used in habitat restoration for riverbank stabilization. The species shown here are sawgrass sedge (*C. spissa*, background and lower right), clustered field sedge (*C. praegracilis*, upper right), and San Luis Obispo sedge (*C. obispoensis*, left), a species found only in this county. *Carex* is Latin for cutter, in reference to the jagged leaf edges and sharply angled stems of some species.

Salt Marsh Dodder
Cuscuta salina* var. *major George Engelmann (1809-1884)
Convolvulaceae *(koos-KOO-tuh sah-LEE-nuh)*

The salt marsh dodder, a member of the morning glory family, is a completely parasitic plant with slender, waxy, orange stems with no chlorophyll and very reduced, non-visible leaves. It receives all its nutrients from twining around green, photosynthetic host plants and inserting little rods, called haustoria, into their tissue in order to steal nourishment. Salt marsh dodder parasitizes a number of plants including, but not limited to, pickleweed (*Sarcicornia pacifica*), fleshy jaumea (*Jaumea carnosa*), and saltbushes (*Atriplex* spp.). There are over ten species of dodder in California that can be found free loading on plants in other habitats. In late spring and summer, dense orange tangles of dodder stems can often be seen from a distance in the salt marsh. In addition to some inland saline flats, this species ranges on the Pacific coast from Baja California north to British Columbia. From May to September, salt marsh dodder produces minute, five-petaled, waxy, white flowers. *Cuscuta* is the Arabic name for dodders, and *salina* is Latin for growing in salty places.

Saltgrass
Distichlis spicata (Linnaeus) Edward Greene (1843-1915)
Poaceae *(diss-TICK-liss spih-KAW-duh)*

Saltgrass is found throughout the United States and Southern Canada as well as in Central and South America. In the central portion of California it grows around salt marshes, along salty beach bluffs, and in dry inland areas with alkaline soil. Like others in these habitats, this is a salt-tolerant (halophytic) species. Some halophytic plants accumulate salt in their tissues (see pickleweed) while others, like salt grass, have specialized glands through which excess salts are excreted as a concentrated liquid, drying on leaf surfaces and forming visible white crystals. This ability to excrete salt has recently made saltgrass an attractive candidate for testing its uses in the reclamation of degraded, excessively salty agricultural soils. It is a mat-forming, dioecious, perennial grass with creeping horizontal underground stems (rhizomes) and stiff angular leaves. These leaves are borne in two vertical ranks along short erect branches, a feature that gives this genus its name. *Distichlis* is derived from *distichos*, a Greek word meaning two-ranked. From April to July saltgrass forms tan to dark purple spike-like floral clusters. *Spicata* means spike-like. Aboriginal Californians harvested a salty dill pickle-flavored seasoning by threshing the leaves.

Alkali Heath
Frankenia salina (Molina) Ivan Johnston (1898-1960)
Frankeniaceae *(fran-KEN-ee-ah sah-LEE-nuh)*

Alkali heath grows in the waterlogged saline soils of coastal salt marshes from Baja California north to Marin County, as well as in alkali flats in dry interior areas. Locally this species can be found in abundance in the tidal salt marshes surrounding Morro Bay. This is a small perennial plant with a slightly woody base and non-woody upper stems (sometimes called a subshrub). The leaves are rolled sharply downward along the margins and are often densely hairy. In the summer, alkali heath produces scattered, small, usually solitary flowers that range in color from burgundy to pink to white. This astringent plant, which is marketed as yerba reuma, is used extensively in modern herbal medicine in the treatment of diseases of the mucous membranes as well as inflammation and pain in the joints, muscles, and fibrous tissue (rheumatism). The genus was named in honor of the sixteenth-century Swedish professor of medicine and botany, Johan Franke, a colleague of Carolus Linnaeus and one of the first people to write about the plants of Sweden. *Salina* in Latin means growing in salty places.

Fleshy Jaumea
Jaumea carnosa (Lessing) Asa Gray (1810-1888)
Asteraceae (*JOW-mee-ah kar-NOE-suh*)

Fleshy jaumea or marsh jaumea is one of only two species in its genus. It is a salt-tolerant, low-growing, perennial member of the sunflower family that forms large mats by growth of horizontal runners and rhizomes. It tolerates seasonal flooding and the poorly drained, salty soils of coastal salt marshes and sea cliff bases where it ranges from Baja California all the way north to Puget Sound and Vancouver Island. This plant's bluish-gray, fleshy, smooth-margined, hairless leaves are arranged oppositely (two per node) and have no petiole. The solitary daisy-like floral heads have yellow ray and disk flowers. This genus was named in honor of Jean Henri Jaume St. Hilaire, a nineteenth-century French botanist who added St. Hilaire to his name later in life. Jaume was supposedly underappreciated professionally and lived in poverty his entire life, even though he published many works on the plants of France. *Carnosa* is from the Latin word *carnosus*; meaning fleshy or pulpy, referring to this species' succulent, salt accumulating leaves.

Rushes

Juncus **spp.** Carolus Linnaeus (1707-1778)
Juncaceae (*JUN-kuss*)

The rushes are members of a large genus (*Juncus*) with nearly 300 species worldwide and over fifty in California. They range from small annuals to large perennials that form dense clumps or spread by rhizomes. They grow in fresh water wetlands, scattered in and around coastal salt marshes, in wet woodlands, and near creeks. Rushes generally have basal, grass-like or needle-like leaves that form a tight sheath around the stem. Some species have floral clusters that appear to emerge laterally from the stem, while in others the floral clusters are obviously terminal. *Juncus* means to join or bind in Latin, in reference to the use of the stems in weaving. The local Chumash Indians used a number of rushes for weaving clothing, mats, and intricately coiled baskets. Both stems and roots were used depending on the desired color. Rush stems were harvested year-round, split longitudinally, and soaked in water before weaving. A number of other wetland plants such at cattails (*Typha* spp.), sedges (*Carex* spp.), and tules (*Schoenoplectus* spp.) were also used extensively in the production of textiles and baskets. The rushes pictured here are spiny rush (*J. acutus* ssp. *leopoldii*, two photos right), brown-headed creeping rush (*J. phaeocephalus* var. *phaeocephalus*, bottom middle), soft rush (*J. effuses*, bottom left), which is grown extensively in Japan for weaving into household flooring mats, salt rush (*J. lesuerii*, top middle), and common rush (*J. patens*, top left).

Sea Lavender

Limonium californicum (Boissier) Amos Heller (1867-1944)
Plumbaginaceae *(lye-MOE-nee-um kal-lih-FOR-ni-kum)*

Limonium is a genus with about 350 species distributed mostly in temperate maritime regions of the world. Sea lavender or western marsh rosemary, the only *Limonium* species native to California, is found on beaches and around the upper margins of tidal marshes along much of the California coast. It is a perennial herb with a woody lower stem and large, leathery, basal leaves that are often rust red colored along their smooth, wavy edges. The tall flowering stems emerge in late summer and fall with branched clusters of pale lavender flowers. Sea lavender's European relatives are cultivated ornamentally for dried cut flowers, which maintain long-lasting color in bouquets. The roots of other related species are used for hide tanning in Russia and medicinally in South America as a form of birth control. *Limonium* comes from the Greek word for a meadow, *leimon*, alluding to its marsh habitat, sometimes called a salt meadow. *Californica*, the species epithet of many plants living on the Central Coast, is used to denote plants that are from California or were first collected here.

Pickleweed

Sarcocornia pacifica Heinrich Schott (1794-1865)
Amaranthaceae *(sar-koe-KORE-nee-ah   pah-SIH-fih-kuh)*

Pickleweed and its close relatives can be found growing on salty, alkaline soils in many parts of the world including the Mediterranean, the eastern seaboard of the U.S., Mexico, and coastal California. Locally, it is abundant in the salty soils of tidal marshes surrounding Morro Bay. This perennial lies flat on the ground, creeping as it grows into enormous clones, with the tips of its jointed succulent stems turned upwards toward the sun. Pickleweed survives the osmotic challenges of living in saltwater by accumulating salt in its succulent tissues. The accrued salt gives the plant, which was eaten raw by aboriginal Californians, a salty taste. In Europe, pickleweed is still pickled in vinegar for a seasoning and was formerly burnt to create soda ash, which was used extensively in glass making (another common name for this species is glasswort). *Sarcocornia* is from the Greek words *sarco*, fleshy, and the Latin word *cornu*, horn, in reference to the appearance of the stems. *Pacifica* means growing near the Pacific Ocean. Until recently, this species went under the scientific name *Salicornia virginica*.

Bulrush or Tule
***Schoenoplectus* spp.** Carol Meyer (1795-1855)
Cyperaceae *(shee-noe-PLEK-tuss)*

Bulrushes are large reed-like plants in the sedge family. They have tall, bright-green, leafless stems that have the shape of a rounded triangle when viewed in cross-section. Inconspicuous, brown flowers and fruit appear in clusters at the tip of these stems. Bulrushes grow around marshes, stream banks, and lakes and have a large native range, living in wetlands across the southern United States, and in temperate regions of South America, as well as on some Pacific islands. These plants covered approximately two million acres of freshwater marshland in the southern San Joaquin Valley prior to the arrival of the first Spaniards, who called these wetland areas tulares. This name, the basis for naming Tulare County, may have its origins in the Spanish word, *tullin*, for cattails (*Typha* spp.). Now only remnants of the once vast tulares exist after being diked, drained, and plowed for agriculture. Aboriginal Californians had many uses for the fibrous stems of bulrushes including baskets, clothing, mats, rafts, and decoys for duck hunting. The species shown above is California bulrush (*S. californicus*). Other commonly encountered species in Central Coast wetlands are the American and common three-squares (*S. americanus* and *S. pungens*). *Schoenoplectus* is from the Greek words *schoinos*, the name for rushes (*Juncus* spp.), and *plectos*, plated or woven.

Arrow-Grass
Triglochin concinna* var. *concinna Burtt Davy (1707-1778)
Juncaginaceae (*try-GLAW-kin conn-SIN-uh*)

Although arrow-grass has the superficial appearance of a grass, it is not a member of the grass family (Poaceae). Instead it is a member of a small plant family of mostly wetland herbs, with only four genera and twenty species, seventeen of which are in the genus *Triglochin*. Arrow-grass is a low-growing, herbaceous perennial that forms lawn-like patches in Pacific coast salt marshes from Baja California to British Columbia. Arrow-grass and pickleweed (*Sarcocornia pacifica*) are often found together, sharing the same tidal belt in the salt marshes surrounding Morro Bay. The cylindrical, arrow-like, slightly fleshy leaves of this species are toxic due to the production and accumulation of cyanide. During spring and summer, greenish-brown flowers are borne on a slender inflorescence type called a raceme, an unbranched, spike-like structure in which the upper flowers open first. The genus name comes from the Greek words, *tri*, three, and *glochis*, a point, referring to the pointed fruit of some species. *Concinna* is Latin for neat, pretty, or well proportioned.

Cattails
Typha spp. Carolus Linnaeus (1707-1778)
Typhaceae (*TYE-fuh*)

Cattails are unmistakable pond, lakeside, roadside ditch, and freshwater marsh dwellers. These extremely common wetland plants can be found throughout California, North America, Europe, Asia, and Africa. They are large perennial herbs that grow tall, flat, parallel-veined leaves from horizontal, underwater rhizomes. One feature that makes these plants so conspicuous is their brown cylindrical, spike inflorescence, which is felt-like to the touch. This inflorescence is made up of thousands of tiny male flowers on the upper portion of the spike and female flowers below, which, after being wind pollinated, develop into minute, wind-dispersed fruits. Aboriginal Californians took full advantage of these incredibly useful plants. In spring, starchy young shoots were cut from rhizomes and eaten like potatoes. Immature flower stalks were boiled and eaten like corn, and the pollen served as nutritious flour. Cattail leaves were woven into flexible baskets, mats, and cordage and the fluffy fruit was used as insulation and stuffing. There are three species of cattails native to California, which all look very similar, broad-leaved cattail (*T. latifolia*, left and right), southern cattail (*T. domingensis*, center), and narrow-leaved cattail (*T. angustifolia*). *Typha* is the ancient Greek name for cattails.

Stinging Nettle
Urtica dioica Carolus Linnaeus (1707-1778)
Urticaceae (*ur-TIH-kuh dye-OY-kuh*)

Stinging nettle is a large perennial herb that spreads by underground rootstocks. It is widely distributed throughout much of the northern hemisphere on stream banks and marsh edges, often forming dense stands in moist places. Its grayish-green, hairy, coarsely-toothed, egg-shaped leaves are arranged opposite each other on the stem. The small, petal-less, greenish flowers are in hanging clusters in the leaf axils. This species is dioecious (as the specific epithet implies), with male and female flowers on different plants. Stinging nettle is covered with hollow, glandular, syringe-like hairs that act as natural hypodermic needles, puncturing the skin and delivering a poisonous concoction containing such compounds as histamine, formic acid, and acetylcholine. Exposure causes a stinging, red, swelling, itching, and burning reaction on the skin. This power to sting remains even in dead plants. Despite this disagreeable feature, stinging nettle still has many virtues. The young shoots are tasty and harmless when boiled, and stem fibers have been used to create fishing line and nets. In addition, both rheumatism and arthritis may be treated by purposefully stinging the affected tissues, a process called urtication. *Urtica* is the ancient Latin name for stinging nettles.

Other salt and freshwater marsh plants. Clockwise from top left. Umbrella sedge (*Cyperus eragrostis*), coastal silver leaf (*Potentilla anserina*), marsh pennywort (*Hydrocotyle verticillata*), and California sea blite (*Suaeda californica*).

Weeds

Ralph Waldo Emerson once described a weed as a plant whose virtues have not yet been discovered. Although this may be true, there are some plants whose virtues are best left to be discovered in their native range. Displaced from their own parts of the world, many weeds are capable of wreaking ecological havoc, as can be found in many areas around San Luis Obispo.

Widespread human travel has helped many plants invade foreign ecosystems, often at the expense and exclusion of native plants. Weedy species have been brought to California, either purposefully or accidentally, since the arrival of the first settlers, and there are now over 1,000 non-native plants that are naturalized (living and reproducing on their own) in the state, with more on the way.

There are several factors contributing to the success of weedy species, one of the most important being their ability to thrive in adverse conditions. After an area is disturbed by human activities, weeds can become established and tenaciously defend their position in a plant community, leaving no opportunity for natives to regrow. The following pages are dedicated to describing some of our most common non-native, naturalized plants.

Wild Mustard

***Brassica* spp.** Carolus Linnaeus (1707-1778)
Brassicaceae (*brass-IH-kuh*)

The wild mustards are widespread, invasive weeds that may have been introduced to California from Europe purposefully by the Spanish Franciscan padres who supposedly scattered seeds so that yellow blooms would mark the Camino Real route. They are annuals with large basal, often deeply lobed leaves, four-petaled, bright yellow flowers, and elongated, cylindrical seedpods often terminating in a beak. They grow in abandoned fields, on open grassy hills, along roadsides, in orchards, and other areas disturbed by human activities. In years of good rainfall, these species are capable of impressive early spring blooms, covering entire areas of rolling coastal hills with unbroken carpets of yellow. The commonly found wild mustards in the San Luis Obispo area are black mustard (*B. nigra*, pictured here), field mustard (*B. rapa*, with upper leaves that clasp around the stem), and two related species now in different genera: charlock (*Sinapis arvensis*, previously *B. kaber*), and the perennial Mediterranean mustard (*Hirschfeldia incana*, previously *B. geniculata*). *Brassica* is the ancient Latin name for cabbage (*B. oleracea*), a species that after generations of selective breeding by humans, has also given us foods like broccoli, cauliflower, brussels sprouts, collard greens, and kohlrabi.

Poison Hemlock
Conium maculatum Carolus Linnaeus (1707-1778)
Apiaceae (*KOE-nee-um   mak-yew-LAY-tum*)

Poison hemlock, a deadly member of the carrot family, is a tall, highly branched, biennial, herbaceous weed with a white, carrot-like taproot. Although originally from Europe, it has now spread throughout North America and become very common in urban areas, along roadsides and stream banks, and in wet places throughout lower altitude regions of California. This plant's maroon-spotted, hollow stems have an offensive odor, like a musty mouse colony. Attached to these stems are fern-like, hairless, highly-divided leaves. Small white flowers are produced in open umbrella-like clusters in the spring and early summer. This is a wickedly poisonous plant that has been implicated in many deaths, the most famous of which is the poisoning of Socrates in Greece. Consuming a small amount of any portion of the plant can cause fatal poisoning, which progresses through successive stages of numbing, convulsions, paralysis, respiratory failure, and finally death. Accidental poisonings still occur regularly because this plant is so widespread and similar to other edible members of the carrot family such as anise, parsley, carrots, fennel, and parsnips. *Conium* comes from *koneion*, the Greek name for the poison derived from this species, the official state poison of ancient Athens, used to execute political prisoners. *Maculatum* is Latin for spotted.

Filaree or Stork's Bill
Erodium **spp.** (Linnaeus) Charles L'Héritier (1746-1800)
Geraniaceae (*ih-ROE-dee-um*)

Filarees are a group of widespread annual weeds native to the Mediterranean. Their adaptability to many different conditions makes them plentiful throughout California in varying habitats. These weeds were most likely purposefully introduced to California shortly after or during the construction of the first Franciscan Missions, as they are considered excellent forage for cattle. Early settlers also ate the fresh stems and leaves raw or cooked. The leaves are either deeply lobed or pinnately compound (leaflets arranged on each side of a common axis) and often form ground level rosettes. Slender flower stalks develop in late winter and early spring, topped with quarter-inch wide magenta or lavender flowers, which then develop into needle-shaped fruits. It is this heron's bill-shaped fruit that gives the genus its name, *Erodium*, which comes from the Greek word for a heron, *erodios*. The fruit breaks apart into one-seeded, corkscrew-shaped segments, which change shape with changing humidity, helping them burrow into the ground, animal fur, or the socks of hikers. Of the seventy-five species of *Erodium*, only eight occur in California. The most common of these are red-stem filaree (*E. cicutarium*, pictured here), green-stem filaree (*E. moschatum*), and long-beak filaree (*E. botrys*).

Blue Gum
Eucalyptus globulus   Jacques de Labillardiere (1755-1834)
Myrtaceae (*yew-kuh-LIP-tuss glawb-YEW-lus*)

The genus *Eucalyptus* is native to Australia, where more than 800 species have evolved in geographic isolation over the past 40 million years. Over 300 species of these beautiful trees have been grown in California and only a few, such as the blue gum, are capable of reproducing in the wild. Native to Southeastern Australia and Tasmania, blue gum is the most common non-native tree in California. Imported to California in the 1850s, first as a horticultural oddity, then as possible salvation for a forecasted timber famine, this fast-growing tree is now naturalized in many areas of the state. Blue gums can become very large trees (the tallest non-conifer tree in California is a blue gum) with long, evergreen, sickle-like leaves, and deciduous bark that sheds seasonally in long strips. The fragrant leaves have oils used in pharmaceuticals, disinfectants, and ointments. The name *Eucalyptus* comes from the Greek words *eu*, meaning well, and *kalyptos*, meaning covered, referring to the flower buds, which are covered by a woody cap that falls off as the flowers open. *Globulus* is Latin for little button, describing the appearance of the mature fruits.

Fennel
Foeniculum vulgare Carolus Linnaeus (1707-1778)
Apiaceae (*foe-NIK-yew-lum vul-GAR-aye*)

Fennel is a native of southern Europe that has escaped from cultivation and is now naturalized in most of the western hemisphere. On the Central Coast, it is a frequently invasive weed that is especially abundant near roadsides, in vacant lots, and on fencerows. Fennel is an erect, highly-branched, perennial herb with stems that have the sent of licorice (*Glycyrrhiza glabra*) or anise (*Pimpinella anisum*). The dark green, finely dissected, thread-like leaves are attached to the stem by long leaf stalks that clasp around the stem. Throughout the summer many tiny, yellow flowers are produced in branched, flat-top umbel clusters, a type of inflorescence seen in most members of the carrot family (Apiaceae). This plant is not to be confused with the similar looking but extremely toxic, poison hemlock (*Conium maculatum*), which has purple spots on its stems, white flowers, and no licorice aroma. Boiled leaf bases of fennel are edible and have a much milder flavor than the upper leaves. Essential oil of fennel is used for flavoring and in traditional Mediterranean medicine to help those who suffer from excessive flatulence. Spanish priests supposedly spread fennel seed on California mission floors to produce a sweet scent under the feet of the congregation. *Foeniculum* was the ancient Latin name for fennel, and *vulgare* means common.

Mallows or Cheeseweeds
Malva **spp.** Carolus Linnaeus (1707-1778)
Malvaceae (*MAL-vuh*)

These herbaceous weeds, which belong to the same plant family as cotton (*Gossypium hirsutum*) and hibiscus (*Hibiscus* spp.), are natives of the Mediterranean and western Asia. Mallows are now naturalized in agricultural fields, gardens, orchards, lawns, vineyards, and on roadsides throughout California. They have very distinct, rounded, shallowly lobed or toothed leaf blades with palmate veins emanating from a reddish center point. The solitary, cup-shaped flowers have five petals ranging in color from white to purple, often with dark veins. After pollination, a small round fruit develops, which tastes a bit like jicama (*Pachyrrhizus erosus*). The resemblance of the fruit to a small cheese wheel is the most likely source of the common name. This wheel splits apart into a number of one-seeded, wedge-shaped segments when the fruit dries. Seeds of many mallows have the amazing ability to remain dormant and persist for many years in the soil without germinating. A study in 1983 showed that seeds of cheeseweed (*M. parviflora*, pictured here) recovered from the adobe bricks used to make a 200-year-old historic California building had retained their ability to germinate. *Malva* is the ancient Greek word for mallows, which may be derived from *malache*, meaning soft, in reference to the leaves.

Bermuda Buttercup
Oxalis pes-caprae Carolus Linnaeus (1707-1778)
Oxalidaceae (*ox-AL-iss pess-KAH-pray*)

The Bermuda buttercup is an insidious urban weed with ostensibly clover-like leaves and beautiful yellow flowers. It isn't from Bermuda, and it isn't a buttercup, nor is it a grass as its other common name, sourgrass, suggests. This species is a perennial South African native that forms small underground bulbs, and flowers through the winter and early spring. The leaves have a pleasant sour taste derived from oxalic acid. Eating a little tastes good; eating a lot can make you ill. Sheep that graze on this plant can develop kidney damage. Bermuda buttercup never sets seed in California, but has managed to spread by its small bulbs in transported soils, on farm machinery, and on muddy shoes. The scientific name is derived from the Greek word *oxys* meaning sour, and the Latin words *pes* and *caprae*, meaning goat's foot, referring to the similarity between a goat's footprint and the shape of this plant's heart-like single leaflets.

Plantain
***Plantago* spp.**   Carolus Linnaeus (1707-1778)
Plantaginaceae (*plan-TAY-go*)

There are over 250 species of *Plantago* in the world, fourteen native or naturalized in California, about half of which occur on the Central Coast. The more conspicuous species have long ribbed leaves and leafless floral spikes terminating in corncob-like clusters of tiny four-petaled flowers. The most common, non-native, naturalized species are English plantain (*P. lanceolata*, pictured in the middle and inset right) and common plantain (*P. major*). Our most widespread native plantain is the minute, annual, California plantain (*P. erecta*, inset left), which often grows no taller than a couple inches. It is found ubiquitously throughout California, flowering in spring. The plantains have numerous medicinal uses, especially in modern Chinese herbalism, and the young leaves of some are edible. *Plantago* means the sole of a foot in Latin, alluding to how the leaves of some lay flat on the ground. Coincidentally, aboriginal Californians called *Plantago* white man's footprint, due to the arrival of these non-native species after every new European incursion.

Castor Bean
Ricinus communis Carolus Linnaeus (1707-1778)
Euphorbiaceae *(RYE-sin-us kom-MEW-niss)*

Castor bean is a native of northeastern Africa that has become naturalized in fields, stream banks, and roadsides in coastal southern and central California. In the San Luis Obispo area, it grows as a robust, perennial shrub with large lobed leaves that range in color from dark purple to bright green. This is a monoecious species with female flowers clustered separately above male flowers on the same inflorescence. Spiny ovaries of female flowers develop into explosive, three-parted capsule fruits, which are capable of ejecting seeds with considerable force while drying. The beautiful and extremely poisonous seeds, which are slightly larger than a kidney bean (although they are botanically unrelated to beans), have a white, black, and brown-spotted body topped with a small spongy caruncle. They superficially resemble an engorged tick and *Ricinus* means tick in Latin. Although in California castor bean is often an overgrown weed, it is widely cultivated throughout the world and has many important economic uses. Most of these uses stem from the thick oil that is pressed from the seeds. This oil is the base for many cosmetics, paints, and varnishes, and is used in the manufacturing of nylon, a number of fruity artificial flavors, and for lubrication in high performance engines. The seeds also contain one of the deadliest natural poisons, called ricin, which is used in cancer chemotherapy. *Communis* is Latin for common.

Curly Dock
Rumex crispus Carolus Linnaeus (1707-1778)
Polygonaceae *(ROO-mex KRISS-pus)*

Curly dock is a perennial weed introduced from Europe that grows in wet places in many different habitats throughout California. It makes a dense rosette of wide, basal, lance-shaped leaves, which have curling, wavy edges. In spring and early summer, a branched inflorescence emerges from the center of the rosette, producing an abundance of tiny green flowers. After pollination, small three-parted fruits develop as the plant turns a deep rust color in the summer. Data from the world's longest continuously monitored experiment has shown curly dock seeds to be very long lived. In 1879, Professor William Beal of Michigan State University buried jars containing fresh seeds from over twenty species, including curly dock. Every five to ten years for the last 120 years one jar has been unearthed and its seeds germinated. Curly dock seeds germinated each time until 1960, more than eighty years after they were originally buried. This plant's carrot-like taproot, which contains anthraquinone glycosides, is used medicinally to treat disorders of the skin, respiratory problems, arthritis, and as a laxative. A crushed leaf applied to the skin can help relieve the pain and itching caused by stinging nettle (*Urtica dioica*) or an insect bite. *Rumex* is the ancient Latin name for docks, and *crispus* means curled.

Weedy Thistles
Asteraceae

Although the four species pictured above share the word thistle in their common names, they are not especially closely related botanically. These ubiquitous, noxious, weedy members of the sunflower family all belong to different genera. Like other members of the sunflower family, their flowers are clustered into compact heads. In most species considered thistles, these flowers are attached to a receptacle that is surrounded by small, scale-like, bristly, spiny, or prickly modified leaves. The four species pictured here (clockwise from top left) are yellow star thistle (*Centaurea solstitialis*), Italian thistle (*Carduus pycnocephalus*), sow thistle (*Sonchus oleraceus*), and milk thistle (*Silybum marianum*). The eminent eighteenth-century botanist Carolus Linnaeus named all four species, which are native to Europe and the Mediterranean. These thistles can be easily found on roadsides, in pastures, and abandoned fields throughout lower elevation portions of California, blooming in the spring and summer. Some species, especially yellow star thistle, have become obnoxious agricultural weeds in California and cause millions of dollars in losses each year.

Other common weeds. Clockwise from top left. Bristly ox tongue (*Picris echioides*), hairy vetch (*Vicia villosa*), pineapple weed (*Chamomilla suaveolens*), tree tobacco (*Nicotiana glauca*), purple star thistle (*Centaurea calcitrapa*), and horehound (*Marrubium vulgare*).

Further Reading

Balls, E. K. (1962). *Early uses of California plants.* Berkeley; University of California Press.

Barbour, M., B. Pavlik, F. Drysdale, and S. Lindstrom (1993). *California's changing landscapes: Diversity and conservation of California vegetation.* Sacramento, California; California Native Plant Society.

Crampton, B. (1974). *Grasses in California.* Berkeley, University of California Press.

Hickman, J. C., et al. (1993). *The Jepson manual: Higher plants of California.* Berkeley; University of California Press.

Holland, V. L. and D. J. Keil (1995). *California vegetation.* Dubuque, Iowa; Kendall/Hunt Publishing Company.

Hoover, R. F. (1970). *The vascular plants of San Luis Obispo County, California.* Berkeley; University of California Press.

Mabberley, D. J. (1997). *The plant-book: A portable dictionary of the vascular plants.* Cambridge; Cambridge University Press.

Ornduff, R., P. M. Faber, and T. Keeler-Wolf (2003). *Introduction to California plant life.* Berkeley; University of California Press.

Raven, P. H., R. F. Evert, and S.E. Eichhorn (2005). *Biology of plants, 7th Edition.* New York, New York; W.H. Freeman and Company/Worth Publishers.

Stearn, W.T. (2002). *Dictionary of plant names for gardeners.* Portland, Oregon; Timber Press

Stuart, J. D. and J. O. Sawyer (2001). *Trees and shrubs of California.* Berkeley; University of California Press.

Line drawings from: von Marilaun, Anton Kerner and F. W. Oliver (1896). *The Natural History of Plants.* London; Blackie and Son.

Glossary

Aboriginal Californian. Peoples present in California prior to European contact.

Achene. A small, single-seeded, dry, indehiscent fruit, for example a sunflower seed.

Allelopathy. The chemical inhibition of one plant by another.

Alternate. A leaf arrangement on a stem in which one leaf is attached per node.

Annual. A non-woody plant that completes its life cycle in one year or growing season.

Anther. The enlarged pollen-forming portion of a stamen borne at the tip of the filament.

Awn. A slender bristle-like appendage borne on a bract of a grass spikelet.

Axil. The upper angle between the stem and the leaf or leaf stalk (petiole).

Basal. Found near the base of a plant, often at ground level, such as leaves clustered near or emanating from the stem at ground level.

Berry. A fleshy, many-seeded, indehiscent fruit.

Biennial. A plant that completes its life cycle in two growing seasons, usually flowering in the second. These plants are usually non-woody, at least above ground.

Bract. A small, modified leaf-like structure associated with reproductive parts, such as inflorescences or flowers.

Bud. An undeveloped shoot or unopened flower.

Bulb. A short, often vertically oriented, underground stem and the fleshy leaves attached it. Bulbs store nutrients.

Bunchgrass. A perennial, clump-forming grass.

Bur. A barbed or bristly fruit or fruit bearing structure that attaches to and is dispersed by the fur of passing animals.

Calyx. The collective term for all the sepals of a flower. The outermost whorl of flower parts, which are typically green.

Capsule. A dry, often many-seeded fruit that splits apart at maturity (dehiscent).

Carpel. The basic female structure of flowering plants in which the ovules are borne. A number of carpels can be fused to form a pistil.

Caruncle. A waxy, oily outgrowth on some seeds.

Catkin. A spike-like, sometimes pendent cluster of flowers that are all the same sex.

Compound leaf. A leaf whose blade is divided into distinct parts (leaflets).

Corolla. The collective term for all the petals of a flower. The whorl of flower parts just inside the calyx.

Cotyledon. A seed leaf on the plant embryo.

Cyme. A cluster of flowers (inflorescence) that terminates with the earliest flower and the central or uppermost flowers open before the outer or lower-most flowers.

Deciduous. A structure on a plant that falls off naturally at the end of a growing period. This is generally said of leaves that fall seasonally or of plants that are seasonally leafless.

Dehiscent. This is said of dry fruits that burst or split open at maturity, for example a pea pod.

Dicot. A member of a main subgroup (a Class) of the flowering plants. These typically have two cotyledons, flowering parts in multiples of four or five, non-parallel leaf venation, and stem veins in rings.

Dioecious. A species in which male and female reproductive structures are produced on separate individuals.

Disk flowers. The small, cylindrical flowers that are clustered to the inside of the head inflorescences made by members of the sunflower family (Asteraceae).

Disturbed. This term describes a weedy, waste area that has been ecologically disrupted or destroyed by human activities or natural processes such as fire.

Dominant. A plant that characterizes a particular habitat or plant community because of its abundance and prevalence.

Drought deciduous. Plants whose leaves fall off during the dry summer.

Drupe. A fleshy fruit with one or more seeds encased in a hard stone, for example a cherry or peach.

Endangered. A species whose survival is in immediate jeopardy.

Endemic. A species that occurs only in a defined area.

Ephemeral. Something that lasts only for a short time.

Evergreen. A plant that has leaves that remain green and attached to the plant and do not fall off all together.

Fascicle. A bunch or tuft of small leaves all arising from the same place.

Filament. The thread-like, stalk portion of a stamen, atop of which sits the anther.

Forb. A non-grassy, herbaceous plant often found in grassland areas.

Frond. A fern leaf, which is often large and lobed or compound.

Fused. Structures that are united together, i.e. all the petals can be fused in a corolla.

Genus. A group of related species and the taxonomic category that ranks above species and below family. Plural of genus is genera.

Glumes. In the grass family (Poaceae), two bracts that enclose the base of a spikelet.

Habitat. A plant's natural environment.

Hair. A thread-like outgrowth of the outermost layer of cells.

Halophyte. A salt tolerant plant.

Haustorium. An outgrowth of a parasitic plant that absorbs nutrients from a host plant.

Head. A dense, spherical cluster of flowers.

Herb. A non-woody plant.

Herbaceous. Having the characteristics of an herb.

Hummock. A mound of plant material and sand found on the outer dunes.

Indehiscent. This is said of fruits that do not open to release their seeds when mature, for example a tomato.

Inflorescence. A cluster of flowers along with any associated structures.

Invasive. A term describing a noxious weed that outcompetes and displaces native species.

Leaflet. A smaller, leaf-like unit of a compound leaf.

Lemma. A bract associated with and positioned directly below a flower (called a floret) in the grass family (Poaceae).

Margin. The edge of a leaf or other organ.

Monocot. A member of a main subgroup (a Class) of the flowering plants. These typically have one cotyledon, flower parts in 3's, parallel leaf veins, and scattered stem veins.

Monoecious. A plant with unisexual male and female flowers separated but still on the same plant.

Native. An indigenous plant that occurs naturally in an area. For California, these are plants that occurred here prior to European contact.

Naturalized. A non-native, weedy plant that reproduces on its own without the help of humans.

Node. The position on the stem to which structures, often leaves, are or were attached.

Nut. A dry, indehiscent fruit with a single seed encased in a hard shell, for example an oak acorn.

Opposite or oppositely arranged. This is said of leaves that arise from the stem in pairs, two per node.

Ovary. The ovule-bearing portion of the pistil, which develops into a fruit.

Ovule. An immature seed.

Palmate. This is said of veins, lobes, or leaflets of a leaf that radiate from a single point, like fingers from the palm of a hand.

Pedicel. The stalk of an individual flower or fruit.

Peduncle. Stalk of an inflorescence or of a flower or fruit borne singly.

Perennial. A plant that lives longer than two years or growing seasons; these plants may or may not have woody above-ground structures. All trees and shrubs are perennial.

Perfoliate. This describes a leaf or pair of leaves that completely wrap around a stem.

Petal. A usually colorful individual member of the corolla.

Petiole. The stalk of the leaf, which connects the leaf blade to the stem.

Phyllary. A modified leaf that surrounds the head inflorescence in the sunflower family (Asteraceae).

Pinnate or pinnately compound. A feather-like compound leaf, with two rows of leaflets on opposite sides of a central axis.

Pinole. An aboriginal Californian food that was a finely ground flour made from the seeds of grasses and forbs.

It was eaten either dry or combined with water to make a mush.

Pistil. The female reproductive structure of a flower, which is composed of the stigma, style, and ovary.

Pith. A soft, spongy tissue that occupies the center of a stem.

Plant community. A grouping of different plant species living together in a particular environment.

Pome. A fleshy fruit in the rose family (Rosaceae), such as apples or pears, where the bulk of the flesh comes from enlarged receptacle or floral parts.

Prickles. Sharp-pointed projections, which are derived from the outer-most layers of cells of a stem or leaf.

Raceme. An unbranched cluster of stalked flowers in which the youngest flowers are at the tip.

Radially symmetrical. An organ, usually a flower, that is divisible into mirror-image halves in three or more ways. A bicycle tire is radially symmetrical.

Rare. A plants whose occurrences are restricted to a small number of areas or individuals.

Ray flowers. In the sunflower family (Asteraceae), the petal-like flowers appearing on the periphery of a head inflorescence, which are accompanied by the more centrally located disk flowers.

Receptacle. The enlarged end of a flower stalk to which flower parts are attached.

Rhizome. A horizontal, underground stem.

Root. The underground structure of a plant that functions in anchorage, absorption and transport of water and nutrients, and food storage.

Rosette. A radiating cluster of leaves at ground level.

Runner. A stem that grows horizontally along the ground.

Seed. A reproductive structure that includes an immature plant (embryo) and a store of food, packaged together within a protective coat.

Sepal. A generally green, individual member of the calyx.

Serpentine. The bluish-green, official California state rock, which is characterized by low levels of calcium and other nutrients and high levels of magnesium, and certain toxic metals such as chromium and nickel.

Sessile. This describes a leaf, flower, or fruit that lacks a stalk and is attached directly to the stem.

Shrub. A relatively short, woody plant with multiple trunks.

Sorus or sori. In many ferns, these are the distinct clusters of spore-forming structures. Often they can be seen as brown spots on the underside of fronds.

Species. A group of organisms that resemble each other closely, are related genetically, and can interbreed. Species is abbreviated sp. (singular), spp. (plural).

Spike. An unbranched, often upright, inflorescence with sessile flowers that usually open from bottom to top.

Spikelet. In the grass family (Poaceae), the smallest unit of the inflorescence. The tiny, wind-pollinated flowers are concealed by overlapping scale-like bracts.

Spine. A sharp-pointed projection derived from a leaf.

Spore. The minute, dispersing, reproductive unit of non-seed plants, such as the ferns or spike-mosses.

Spur. A hollow, floral projection that usually contains nectar.

Stamen. The male reproductive structure of a flower, which is composed of a stalk-like filament and a terminal, pollen-producing anther.

Stem. The aboveground (sometimes belowground) axis of a plant.

Stigma. The terminal, often sticky, portion of a pistil on which pollen is deposited.

Strobilus. A cone or cone-like structure.

Style. The stalk-like section of the pistil that connects the ovary to the stigma.

Subshrub. A small shrub with woody lower stems and non-woody upper stems that die back seasonally.

Subspecies. A geographically and often physically distinct population within a species, abbreviated ssp.

Subtending. Occurring immediately below. A bract may subtend a flower.

Succulent. A plant or plant organ that is especially fleshy and usually drought-tolerant.

Tendril. A plant stem or leaf that is modified as a delicate, commonly twisted, appendage that helps the plant climb.

Terminal. At the end or tip of a structure.

Thorn. A sharply pointed branch.

Tooth. A small, pointed projection on the edge of a leaf.

Tree. A tall, woody plant, usually with one large trunk.

Umbel. An umbrella-shaped inflorescence with flowers and pedicels radiating from a single point.

Vine. A trailing or climbing plant that can be woody or herbaceous.

Whorl. A group of three or more leaves, floral parts, or flowers that are clustered at a single node or point.

Woody. A portion of a plant, usually a perennial shrub or tree, that is hard, thickened, and generally covered in bark.

Index

A

Abronia latifolia 71
Abronia maritima 71
Abronia umbellata 71
absinthe 6
achene 30, 107
Achillea millefolium 2, 16
Achilles 2
acorns 51
Adenostoma fasciculatum 5
Adiantum jordanii 39
agriculture 144
algae 66
alkali heath 122
alkaloids 7
allelopathy 137
Ambrosia chamissonis 73
American three-square 127
Amsinckia menziesii 86
Amsinckia spectabilis 86
Anemopsis californica 118
anise 135, 138
anthracnose 45
anthraquinone glycosides 143
Aquilegia eximia 67
Arbutus menziesii 21
Arctostaphylos 3
Arctostaphylos morroensis 3
Aristotle 108
arrow-grass 128
arroyo lupine 103
arroyo willow 59
Artemisia 6, 41
Artemisia absinthium 6
Artemisia californica 6
Artemisia douglasiana 41
Asclepias eriocarpa 112
aspirin 59
Asteraceae 100
Astragalus curtipes 7
Astragalus nuttalii 7
Atriplex 117, 120
Atriplex californica 117
Atriplex lentiformis ssp. *lentiformis* 117
Atriplex triangularis 117
Atriplex watsonii 117
Australia 137
Avena barbata 87
Avena fatua 87
Avena sativa 87
avocados 65
awn 88

B

baby blue-eyes 105
Baccharis pilularis 8
bacteria 24, 66, 103
barley 97
bay-laurel 65
beach, the 69
beach-bur sage 73
beach morning glory 82
beach primrose 75
beach sand verbena 71
Beal, William 143
bedstraw 43
Bentham, George 31, 107
Bermuda buttercup 140
Betula 11
big saltbush 117
birch 11
blackberry 58
black mustard 134
black sage 32
Blochman's leafy daisy 82
Bloomeria crocea 112
blowouts 69
blue-eyed grass 108
blue dicks 94
blue elderberry 33
blue gum 137
blue oak 51
bracken fern 39
Brassica geniculata 134
Brassica kaber 134
Brassica oleracea 134
Brassica rapa 134
bristly ox tongue 145
broad-leaved cattail 129
broccoli 134
brome 88
Bromus diandrus 88
Bromus madritensis 88
broom 24
brown-headed creeping rush 124
brussels sprouts 134
buckwheat 14, 81

buck brush 10
bulbs 92, 140
bulrush 127
bunch grasses 85
bush poppy 35
buttercup 107, 140
butterfly mariposa 89

C

cabbage 134
Cakile maritima 74
California bay-laurel 65
California buckwheat 14
California bulrush 127
California buttercup 107
California coffeeberry 54
California croton 78
California everlasting 29
California figwort 67
California floristic province 49
California hedge nettle 62
California honeysuckle 23
California man-root 25
California Native Plant Society 89
California peony 27
California plantain 141
California polypody fern 39
California poppy 96
California sagebrush 6
California saltbush 117
California sea blite 131
California sycamore 45
California wild rose 30
californica 54
Calochortus albus 89
Calochortus clavatus 89
Calochortus obispoensis 89
Calochortus venustus 89
Calystegia macrostegia 9
Calystegia soldenella 82
Camino Real 134
Camissonia cheiranthifolia 75
camphor 65
cancer 39
Carduus pycnocephalus 144
Carex 119, 124
Carex obispoensis 119
Carex praegracilis 119
Carex spissa 119
Carpobrotus 76, 77
Carpobrotus chilensis 76

Carpobrotus edulis 76
Carrizo Plain 85
carrots 135
carrot family 135
caruncle 78, 142
Castilleja affinis 90
Castilleja affinis ssp. *affinis* 90
Castilleja exserta 90, 91
castor bean 142
catkins 59
cattails 124, 127, 129
cauliflower 134
Ceanothus cuneatus 10
Ceanothus pappillosus 35
Centaurea calcitrapa 145
Centaurea solstitialis 144
Cercocarpus betuloides 11
chamise 5
Chamisso, Adelbert von 73, 75, 96
Chamomilla suaveolens 145
chaparral 1, 3, 5
chaparral honeysuckle 23
chaparral pea 35
chaparral Yucca 19
chaparro 5
charlock 134
checker bloom 112
cheese 43
cheeseweeds 139
cheese making 43
Cheiranthus 75
cherry 28, 55
chia 31
chinese houses 112
Chlorogalum pomeridianum 92
chocolate lily 95
Christmas berry 21
Chumash 8, 49, 124
Cinchona 43
cinnamon 65
Cirsium occidentale 82
Clark, William 93
Clarkia bottae 93
Clarkia purpurea ssp. *viminea* 93
Clarkia unguiculata 93
Clayton, John 42
Claytonia perfoliata 42
Cleveland, Daniel 95
climbing bedstraw 43
climbing penstemon 22
clovers 109
clustered field sedge 119

coastal buckwheat 81
coastal redwood 105
coastal scrub 1
coastal silver leaf 131
coast live oak 49
coast live oak woodlands 37
cob web thistle 82
coffee 43
coffeeberry 54
coffee fern 39
collard greens 134
Collinsia heterophylla 112
columbine 67
common fiddleneck 86
common monkey flower 44
common phacelia 106
common rush 124
common snowberry 63
common three-square 127
Conicosia pugioniformis 77
Conium maculatum 135, 138
corm 94
coyote brush 8
creeping snowberry 63
crisp dune mint 82
Croton californicus 78
crustose lichen 66
cudweeds 29
cultivated oats 87
curly dock 143
Cuscuta salina var. *major* 120
cyme 106
Cyperus eragrostis 131

D

deer weed 24
Deinandra fasciculata 112
Delphinium parryi 94
Dendromecon rigida 35
Dichelostemma capitatum 94
digger pine 47
disk flowers 100
Distichlis spicata 121
diviner's sage 60
dock 143
dodder 120
Dodecatheon clevelandii 95
Don, David 47
Douglas, David 34, 41, 51, 105
drought deciduous 1, 6
drupes 28

Dudley, William Russell 13
Dudleya 13
Dudleya abramsii 13
Dudleya caespitosa 13
Dudleya lanceolata 13
Dudleya pulverulenta 13
dunedelion 82
dunes 69
dune scrub 69, 80
dune wallflower 82
dwarf clover 109

E

ecological invasion 85
Ehrhart, J. Friedrich 79
Ehrharta calycina 79
elderberry 33
elegant clarkia 93
Emerson, Ralph Waldo 133
England 41
English oak 53
English plantain 141
Equisetum hyemale 40
Equisetum telmateia ssp. *braunii* 40
Erica 80
Ericameria 80
Ericameria ericoides 80
Erigeron blochmaniae 82
Eriogonum 14, 81
Eriogonum fasciculatum 14
Eriogonum parvifolium 81
Eriophyllum confertiflorum 16
Eriophyllum staechadifolium 17
Erodium 136
Erodium botrys 136
Erodium cicutarium 136
Erodium moschatum 136
erosion control 79
Erysimum insulare ssp. *suffrutescens* 82
Eschscholtz, J. F. 96
Eschscholzia californica 96
Eucalyptus 137
Eucalyptus globulus 137
everlastings 29

F

Fagopyrum esculentum 14
fascicles 5
felt-leaved everlasting 29
fennel 135, 138
ferns 39

fiddlehead 39
fiddleneck 86
field mustard 134
figwort 67
filarees 136
fire 1, 3, 5, 10
firefighters 64
flannelbush 35
fleshy jaumea 120, 123
floristic province 49
Foeniculum vulgare 138
foliose 66
foothill pine 47
four spot 93
foxtail barley 97
Fragaria chiloensis 82
Franciscan Missions 49, 136
Franciscan Padres 134
Franke, Johan 122
Frankenia salina 122
Franseria 73
Fremontodendron californicum 35
Freshwater marshes 115, 127
Fritillaria biflora 95
fronds 39
fruticose lichen 66
fuchsia-flowered gooseberry 57
fungi 45, 66

G

Galium andrewsii 43
Galium aparine 43
Galium porrigens 43
Galium vernum 43
gardenias 43
giant horsetail 40
Gilia tenuiflora 112
Glycyrrhiza glabra 138
Gnaphalium 29
godetia 93
goldback fern 39
goldenbush 20
golden stars 112
golden yarrow 16
goldfields 99
gooseberry 57
goose grass 43
grasslands 85
Gray, Asa 110
gray pine 47
grazing 87

Greek Gods 27
Greene, Edward 20, 73

H

hairy vetch 145
halophytes 115, 121
Haplopappus 20
haustoria 120
Hawaii 41
hayfield tarweed 112
Hazard, Barclay 20
Hazardia squarrosa 20
heart-leaved penstemon 22
heath 122
heather 80
hedgenettle 62
Hemizonia congesta ssp. luzulifolia 112
hemlock 135
Hesperoyucca whipplei 19
Heteromeles arbutifolia 21
Hilo, Hawaii 41
hip, rose 30
Hirschfeldia incana 134
Hoita macrostachya 67
hollies 21, 28
holly-leafed cherry 28
Hollywood 21
honeysuckle 23
hop clover 109
Hordeum murinum 97
Hordeum vulgare 97
horehound 145
Horkelia cuneata 82
horsetails 40
Hottentot fig 76
hummingbird sage 60
hummock 69
Hydrocotyle verticillata 131

I

Ice Plant 76, 77
imbricate phacelia 106
Indian paintbrush 90
Indian pink 82
invasive grasses 85
Iris Family 108
Italian ryegrass 101
Italian thistle 144
ivy 64

J

Japan 124
Japanese honeysuckle 23
jaumea 123
Jaumea carnosa 120
Jaume St. Hilaire, Jean Henri 123
Jefferson, Thomas 93
Jepson, Willis Lynn 20, 80
Jepson Manual, The 80
Jersey cudweed 29
Jesus 43
jewelflower 112
jicama 139
Juncus acutus ssp. *leopoldii* 124
Juncus effuses 124
Juncus lesuerii 124
Juncus patens 124
Juncus phaeocephalus var. *phaeocephalus* 124

K

Keck, David 22
Keckia 22
Keckiella 22
Keckiella cordifolia 22
kohlrabi 134

L

larkspur 94
Lasthenia californica 99
lavender 125
Lay, George T. 100
Layia platyglossa 100
leather oak 35
leather root 67
lemma 87
leopard lily 67
Lessing, Christian 73
Lewis and Clark 93
lichens 66
licorice 138
lilies 67, 89, 95
Lilium pardalinum ssp. *pardalinum* 67
Limonium californicum 125
Linanthus californicus 35
Linnaeus, Carolus 2, 8, 73, 144
Linnean Natural History Society 47
livestock forage 79
live forever 13
lizard tail 17
locoweed 7

Lolium multiflorum 101
London plane tree 45
long-beak filaree 136
Lonicera hispidula 23
Lonicera interrupta 23
Lonicera involucrata var. *ledebourii* 23
Lonicera japonica 23
Lonicera subspicata 23
Lonitzer, Adam 23
Lotus scoparius 24
Louisiana Purchase 93
Lupines 103
Lupinus albifrons 103
Lupinus chamissonis 103
Lupinus hirsutissimus 103
Lupinus nanus 103
Lupinus succulentus 103

M

Madrid 88
madrone 21
maiden-hair fern 39
Malacothrix incana 82
malaria 43
mallows 139
Malva parviflora 139
man-root 25
mangos 64
Manzanitas 3
Marah fabaceus 25
mariposa lilies 89
Marrubium vulgare 145
marshes 115
marsh jaumea 123
marsh pennywort 131
Mediterranean mustard 134
Melica imperfecta 112
melic grass 112
Menzies, Archibald 105
milk thistle 144
Mimulus 26, 44
Mimulus aurantiacus 26
Mimulus guttatus 44
Miner's Lettuce 42
mint 82
missions 136
Mississippi 93
Miwok Indians 17
mock heather 80
Monardella crispa 82
monkey flower 26, 44

morning glory 9, 82
Morro Bay 115, 122, 126, 128
moss 61
mountain mahogany 11
mugwort 41
mustard 134

N

narrow-leaved cattail 129
narrow leaf clover 109
Nassella cernua 104
Nassella lepida 104
Nassella pulchra 104
naturalized plants 133
Neé, Luis 53
Nemophila menziesii 105
nettle 62, 130
neurotoxins 2
Nicotiana glauca 145
nightshades 34
Nipomo-Guadalupe Dunes 77
nitrogen 24
nitrogen fixing bacteria 24, 103
non-native grasses 85

O

Oakland 49
oaks 37, 49, 51, 53, 65
oak woodlands 37
oats 87
oil 142
Orthocarpus 91
our lord's candle 19
oxalic acid 140
Oxalis pes-caprae 140

P

Pachyrrhizus erosus 139
Pacific madrone 21
Paeon 27
Paeonia californica 27
paintbrush 90
Parry's larkspur 94
parsley 135
parsnips 135
Paso Robles 53
Pellaea andromedifolia 39
Penstemon 22
Pentagramma triangularis 39
peony 27
perennial bunchgrass 104

Phacelia brachyloba 106
Phacelia distans 106
Phacelia grisea 106
Phacelia imbricata 106
phloem 61
Pholisma arenaria 82
phyllaries 29
Pickeringia montana 35
pickleweed 120, 126, 128
Picris echioides 145
Pimpinella anisum 138
pines 47
pineapple weed 145
pine cones 47
pine needles 47
pink cudweed 29
pinole 17, 93
pinole clover 109
Pinus sabiniana 47
pistachios 64
Plantago erecta 141
Plantago lanceolata 141
Plantago major 141
plantain 141
Platanus racemosa 45
Platanus x acerifolia 45
Plato 99
Pliny 95
Poison Hemlock 135, 138
poison ivy 64
poison oak 8, 41, 58, 64
Polypodium californicum 39
polypody fern 39
pome 21
poppy 35, 96
Potentilla anserina 131
prickly bedstraw 43
prickly phlox 35
primrose 75
Prunus ilicifolia 28
Pseudognaphalium beneolens 29
Pseudognaphalium californicum 29
Pseudognaphalium luteo-album 29
Pseudognaphalium ramosissimum 29
Pteridium aquilinum 39
punch-bowl godetia 93
purple needle grass 104
purple nightshade 34
purple owl's clover 90, 91
purple sand food 82
purple sand verbena 71
purple star thistle 145

Q

Quercus 49, 51, 53
Quercus agrifolia 37, 49, 51, 53
Quercus douglasii 47, 51
Quercus durata 35
Quercus lobata 53
Quercus robur 53
quinine 43

R

rabbitfoot clover 109
railroad survey 61
ranunculin 107
Ranunculus californicus 107
rattleweed 7
ray flowers 100
receptacle 144
red-stem filaree 136
redberry 55
redwood 105
red brome 88
resurrection plant 61
Rhamnus 54, 55
Rhamnus californica 54
Rhamnus crocea 55
rheumatism 122
Rheum x hybridum 14
rhubarb 14
Ribes 57
Ribes speciosum 57
ricin 142
Ricinus communis 142
riparian 37
ripgut brome 88
Robles 53
Rosa californica 30
rose 30
rosettes 136
rose hips 30
Royal Horticulture Society 41, 105
Rubus ursinus 58
Rumex crispus 143
Rurick 75
rushes 124
ryegrass 101

S

sagebrush 6
sages 6, 31, 32, 41, 60
Salicornia virginica 126
salicylic acid 59
Salix lasiolepis 59
saltbushes 117, 120
saltgrass 121
saltwater marshes 115
salt accumulation 121, 123, 126
salt excretion 121
salt marsh dodder 120
salt rush 124
Salvia 31, 32
Salvia carduacea 112
Salvia columbariae 31
Salvia divinorum 60
Salvia mellifera 32
Salvia spathacea 60
Sambucus mexicana 33
sand strawberry 82
sand verbena 71
San Diego Natural History Society 95
San Joaquin Valley 127
San Luis mariposa 89
San Luis Obispo sedge 119
San Luis star-tulip 89
Sarcicornia pacifica 120, 126
saw-toothed goldenbush 20
sawgrass sedge 119
Scabiosa columbaria 31
Schoenoplectus 124, 127
Schoenoplectus americanus 127
Schoenoplectus californicus 127
Schoenoplectus pungens 127
Schwarzenegger, Arnold 104
scouring rushes 40
Scrophularia californica 67
scrub oaks 5
seaside fiddleneck 86
seaside woolly sunflower 17
sea blite 131
sea fig 76
sea lavender 125
sea rocket 74
sedges 119, 124
sedge family 127
Selaginella bigelovii 61
selenium 7
Sequoia sempervirens 105
serpentine 13
shaman 60
shell creek 85
shooting star 95
Shoshone Indians 23
shrubs 33

Index | 157

Sidalcea malviflora 112
Sierra Club 13
Silene laciniata 82
silver dune lupine 103
silver lupine 103
Silybum marianum 144
Sinapis arvensis 134
Sisyrinchium bellum 108
sky lupine 103
Slender-leaved Ice Plant 77
slender tarweed 112
slender wild oats 87
snowberry 63
soap plant 92
Socrates 135
soft rush 124
Solanum douglasii 34
Solanum xanti 34
Sonchus oleraceus 144
sori 39
southern cattail 129
southern honeysuckle 23
sow thistle 144
spike moss 61
spiny rush 124
spores 39, 40
spur 94
stabilized dunes 69
Stachys bullata 62
Stanford 13
star-tulip 89
star lily 112
sticky monkey flower 26
stinging lupine 103
stinging nettle 62, 130, 143
storksbill 136
Straw, Richard 22
Streptanthus glandulosus 112
strobili 40
Stylomecon heterophylla 112
Suaeda californica 131
succulent 13, 126
sunflower family 100
Sweden 122
sweet potato 9
sycamore 45, 65
symbiont 66
Symphoricarpos albus var. *laevigatus* 63
Symphoricarpos mollis 63

T

tarweeds 112
tattoos 34
Tegeticula maculata 19
Theophrastus 108
Thousand Oaks 49
thujone 2, 6
ticks 78
Tidy Tips 100
Torrey, John 110
Toxicodendron diversilobum 8, 41, 58, 64
Toxicoscordion fremontii 112
toyon 21
trees vs. shrubs 33
tree tobacco 145
Trifolium angustifolium 109
Trifolium arvense 109
Trifolium bifidum 109
Trifolium campestre 109
Trifolium depauperatum 109
Trifolium wormskioldii 109
Triglochin concinna var. *concinna* 128
tulares 127
Tulare County 127
Tules 124, 127
twinberry 23
Typha 124, 127, 129
Typha angustifolia 129
Typha domingensis 129
Typha latifolia 129

U

University of California, Berkeley 80
Umbellularia californica 65
urtication 130
Urtica dioica 62, 130, 143
urushiol 64

V

valley oak 53
Vancouver, George 105
Vandenberg Air Force Base 77
Van Gogh 6
vascular tissue 61
veldt grass 79
Verbena 71
vernal pools 99
Vicia villosa 145
Virgil 101
Virginia 42

W

wart leaf ceanothus 35
Watson's saltbush 117
weeds 133
weedy grasses 85
western marsh rosemary 125
wetlands 115
Whipple, A.W. 19
white fairy lantern 89
wildflowers 85
wild blackberry 58
wild morning glory 9
wild mustard 134
wild oats 87
wild rose 30
willows 59
windowed phacelia 106
wind poppy 112
woodlands 37
Wormskjold's clover 109

X

Xantus, John 34
xylem 61

Y

yarrow 2, 16
yellow mariposa 89
yellow sand verbena 71
yellow star thistle 144
yerba mansa 118
yerba reuma 122
yucca 19
yucca moth 19
Yuki Indians 23